NIST Technical Note 1734

Airflow and Indoor Air Quality Models of DOE Reference Commercial Buildings

Lisa C. Ng
Andrew K. Persily
Steven J. Emmerich
Energy and Environment Division
Engineering Laboratory

Amy Musser
Vandemusser Design PLLC

February 2012

U.S. Department of Commerce
John E. Bryson, Secretary

National Institute of Standards and Technology
Patrick D. Gallagher, Under Secretary of Commerce for Standards and Technology and Director

ABSTRACT

Sixteen reference buildings have been defined by the U.S. Department of Energy, and created as EnergyPlus input files, for use in assessing new technologies and supporting the development of energy codes in pursuing building energy efficiency improvements. Infiltration rates in the EnergyPlus models of the reference buildings were input as constant airflow rates, and not calculated based on established building airflow theory. In order to support more physically-based airflow calculations, as well as indoor air quality analysis, models of the 16 reference buildings were created in the multizone airflow and contaminant transport program CONTAM. A number of additional inputs had to be defined for the CONTAM models, and changes in the interior zoning were required, to more realistically account for airflow. Annual airflow and contaminant simulations were performed in CONTAM for six of the buildings. While the assumed infiltration rates in EnergyPlus do not realistically reflect impacts of weather conditions, there are clear relationships between the outdoor air change rates calculated by CONTAM and weather. In addition, the envelope airtightness values assumed in either approach are seen to have a major impact on the air change rates. Contaminant analyses were performed for occupant-generated carbon dioxide, volatile organic compounds from indoor sources, outdoor particulate matter, and outdoor ozone. The airflow and contaminant calculation results provide a useful baseline for subsequent use of these models to investigate approaches to building ventilation and other technologies that are intended to simultaneously reduce building energy consumption while maintaining or improving indoor air quality.

Keywords: airflow, energy, CONTAM, EnergyPlus, IAQ, reference buildings, ventilation

TABLE OF CONTENTS

LIST OF TABLES v

LIST OF FIGURES vii

1. INTRODUCTION 1

2. BUILDING DESCRIPTIONS 3

3. MODELING APPROACH 3

    3.1. CONTAM model inputs 4

    3.2. EnergyPlus model inputs 8

4. AIRFLOW SIMULATION RESULTS 10

    4.1. Outdoor air change rates 10

    4.2. Outdoor air change rates vs. weather conditions 19

    4.3. Impacts of infiltration on sensible load 27

5. CONTAMINANT SIMULATION RESULTS 29

6. DISCUSSION 52

    6.1. Building models for airflow and energy analyses 52

    6.2. Limitations of study 54

    6.3. Presenting IAQ simulation results 54

    6.4. Future work 55

7. CONCLUSION 56

8. REFERENCES 57

Appendix A Detailed description of CONTAM models of reference buildings 61

  A1. INTRODUCTION 61

  A2. BUILDINGS DESCRIPTION 63

    A2.1. Quick Service Restaurant 63

    A2.2. Full Service Restaurant 65

    A2.3. Small Office 67

    A2.4. Medium Office 70

    A2.5. Large Office 74

    A2.6. Primary School 78

    A2.7. Secondary School 84

    A2.8. Stand-Alone Retail 92

    A2.9. Strip Mall 95

A2.10.   Supermarket 97

A2.11.   Small Hotel 100

A2.12.   Large Hotel 106

A2.13.   Hospital 113

A2.14.   Outpatient Health Care 123

A2.15.   Warehouse 134

A2.16.   Midrise Apartment 137

Appendix B Detailed calculated contaminant concentration predictions 141

B1.   Full Service Restaurant 141

B2.   Hospital 142

B3.   Medium Office 144

B4.   Primary School 146

B5.   Small Hotel 147

B6.   Stand-Alone Retail 149

LIST OF TABLES

Table 1 Summary of reference buildings 4

Table 2 Properties of contaminants simulated in CONTAM 7

Table 3 Summary of outdoor contaminant concentrations for Chicago 7

Table 4 Buildings for which economizer modeled for at least one HVAC system (Chicago) 9

Table 5 Summary of calculated outdoor air change rates 12

Table 6 Sensible loads due to infiltration 27

Table 7 Selected zones for which contaminant concentration results reported 30

Table 8 Summary of calculated contaminant concentrations 31

Table A1 Summary of zones in Quick Service Restaurant 63

Table A2 Summary of HVAC system flow rates (m^3/s) in Quick Service Restaurant 64

Table A3 Summary of zones in Full Service Restaurant 65

Table A4 Summary of HVAC system flow rates (m^3/s) in Full Service Restaurant 66

Table A5 Summary of zones in Small Office 68

Table A6 Summary of HVAC system flow rates (m^3/s) in Small Office 69

Table A7 Summary of zones in Medium Office 70

Table A8 Summary of VAV system flow rates (m^3/s) in Medium Office for
New and Post-1980 buildings 73

Table A9 Summary of CAV system flow rates in Medium Office for Pre-1980 building 73

Table A10 Summary of zones in Large Office 75

Table A11 Summary of VAV system flow rates (m^3/s) in Large Office 77

Table A12 Summary of zones in Primary School 79

Table A13 Summary of HVAC system flow rates (m^3/s) in Primary School 82

Table A14 Summary of zones in Secondary School 85

Table A15 Summary of HVAC system flow rates (m^3/s) in Secondary School 89

Table A16 Summary of zones in Stand-Alone Retail 92

Table A17 Summary of HVAC system flow rates (m^3/s) in Stand-Alone Retail 94

Table A18 Summary of zones in Strip Mall 95

Table A19 Summary of HVAC system flow rates (m^3/s) in Strip Mall 96

Table A20 Summary of zones in Supermarket 97

Table A21 Summary of HVAC system flow rates (m^3/s) in Supermarket 99

Table A22 Summary of zones in Small Hotel 101

Table A23 Summary of HVAC system flow rates (m^3/s) in Small Hotel 103

Table A24 Summary of PTAC flow rates (m^3/s) in Small Hotel 104

Table A25 Summary of zones in Large Hotel 106

Table A26 Summary of VAV system flow rates (m^3/s) in Large Hotel for
New and Post-1980 buildings 110

Table A27 Summary of CAV system flow rates (m^3/s) in Large Hotel for Pre-1980 buildings 110

Table A28 Summary of DOAS flow rates (m^3/s) in Large Hotel for all building vintages 111

Table A29 Summary of zones in Hospital 114

Table A30 Summary of VAV and CAV system flow rates (m^3/s) in Hospital 121

Table A31 Summary of zones in Outpatient Health Care 123

Table A32 Summary of VAV system flow rates (m^3/s) in Outpatient Healthcare 131

Table A33 Summary of zones in Warehouse 134

Table A34 Summary of CAV system flow rates (m^3/s) in Warehouse 136

Table A35 Summary of zones and outside air rates in Midrise Apartment 137

LIST OF FIGURES

Figure 1 Wind pressure profile simulated in CONTAM for exterior walls.................................... 5

Figure 2 Frequency distribution of simulated outdoor air change rates for
Full Service Restaurant 16

Figure 3 Frequency distribution of simulated outdoor air change rates for Hospital 16

Figure 4 Frequency distribution of simulated outdoor air change rates for Medium Office 17

Figure 5 Frequency distribution of simulated outdoor air change rates for Primary School........ 17

Figure 6 Frequency distribution of simulated outdoor air change rates for Small Hotel 18

Figure 7 Frequency distribution of simulated outdoor air change rates for Stand-Alone Retail.. 18

Figure 8 Air change rates as a function of temperature difference (low wind speed) for
Full Service Restaurant 21

Figure 9 Air change rates as a function of temperature difference (low wind speed) for
Hospital 21

Figure 10 Air change rates as a function of temperature difference (low wind speed) for
Medium Office.. 22

Figure 11 Air change rates as a function of temperature difference (low wind speed) for
Primary School... 22

Figure 12 Air change rates as a function of temperature difference (low wind speed) for
Small Hotel 23

Figure 13 Air change rates as a function of temperature difference (low wind speed) for
Stand-Alone Retail... 23

Figure 14 Air change rates as a function of wind speed (low ΔT) for Full Service Restaurant ... 24

Figure 15 Air change rates as a function of wind speed (low ΔT) for Hospital 24

Figure 16 Air change rates as a function of wind speed (low ΔT) for Medium Office................ 25

Figure 17 Air change rates as a function of wind speed (low ΔT) for Primary School............... 25

Figure 18 Air change rates as a function of wind speed (low ΔT) for Small Hotel 26

Figure 19 Air change rates as a function of wind speed (low ΔT) for Stand-Alone Retail.......... 26

Figure 20 Frequency distribution of simulated CO_2 concentration for Full Service Restaurant.. 34

Figure 21 Frequency distribution of simulated CO_2 concentration for Hospital 35

Figure 22 Frequency distribution of simulated CO_2 concentration for Medium Office.............. 36

Figure 23 Frequency distribution of simulated CO_2 concentration for Primary School.............. 37

Figure 24 Frequency distribution of simulated CO_2 concentration for Small Hotel 38

Figure 25 Frequency distribution of simulated CO_2 concentration for Stand-Alone Retail......... 39

Figure 26 Frequency distribution of simulated VOC concentration for Full Service Restaurant 40

Figure 27 Frequency distribution of simulated VOC concentration for Hospital 41

Figure 28 Frequency distribution of simulated VOC concentration for Medium Office 42

Figure 29 Frequency distribution of simulated VOC concentration for Primary School 43

Figure 30 Frequency distribution of simulated VOC concentration for Small Hotel 44

Figure 31 Frequency distribution of simulated VOC concentration for Stand-Alone Retail 45

Figure 32 Frequency distributions of simulated ozone and PM2.5 daily average
        concentrations for Full Service Restaurant 46

Figure 33 Frequency distributions of simulated ozone and PM2.5 daily average
        concentrations for Hospital 47

Figure 34 Frequency distributions of simulated ozone and PM2.5 daily average
        concentrations for Medium Office 48

Figure 35 Frequency distributions of simulated ozone and PM2.5 daily average
        concentrations for Primary School 49

Figure 36 Frequency distributions of simulated ozone and PM2.5 daily average
        concentrations for Small Hotel 50

Figure 37 Frequency distributions of simulated ozone and PM2.5 daily average
        concentrations for Stand-Alone Retail 51

Figure A1 Floor plan of Quick Service Restaurant (height 3.05 m) 63

Figure A2 Occupancy schedule for Quick Service Restaurant 65

Figure A3 Floor plan of Full Service Restaurant (height 3.05 m) 66

Figure A4 Occupancy schedule for Full Service Restaurant 67

Figure A5 Floor plan of Small Office (height 3.05 m) 68

Figure A6 Occupancy schedule for Small Office 69

Figure A7 Floor plan of Medium Office (height 2.74 m) 70

Figure A8 Occupancy schedule for Medium Office 74

Figure A9 First floor plan of Large Office (height 2.74 m). Second through twelfth floors
        are identical to first floor 75

Figure A10 Occupancy schedule for Large Office 78

Figure A11 Plan of Primary School (height 4.0 m) 80

Figure A12 Occupancy schedules for Primary School (Gym, Cafeteria) 83

Figure A13 Occupancy schedules for Primary School (Offices, Class) 83

Figure A14 First floor plan of Secondary School (height 4.0 m) 86

Figure A15 Second floor plan of Secondary School (height 4.0 m) 87

Figure A16 Occupancy schedules for Secondary School (Gym, Cafeteria, and Auditorium) 91

Figure A17 Occupancy schedules for Secondary School (Offices, Class) 92

Figure A18 Floor plan of Stand-Alone Retail (height 6.1 m) 93

Figure A19 Occupancy schedule for Stand-Alone Retail 94

Figure A20 Floor plan of Strip Mall (height 5.18 m) 95

Figure A21 Occupancy schedule for Strip Mall 97

Figure A22 Floor plan of Supermarket (height 6.1 m) 98

Figure A23 Occupancy schedule for Supermarket 100

Figure A24 (a) First and (b) upper floor (2-4) plans of Small Hotel 102

Figure A25 Occupancy schedules for Small Hotel (Restroom and Exercise) 104

Figure A26 Occupancy schedules for Small Hotel (Lounges, Laundry, Meeting Room,
    Office) 105

Figure A27 Occupancy schedule for Small Hotel (Guest) 105

Figure A28 (a) First, (b) second to fifth, and (c) sixth floors plans of Large Hotel 108

Figure A29 Occupancy schedules for Large Hotel (Lobby, Guest) 112

Figure A30 Occupancy schedule for Large Hotel (Building) 112

Figure A31 First floor plan of Hospital (height 4.27 m), all dimensions in meters 116

Figure A32 Second floor plan of Hospital (height 4.27 m), all dimensions in meters 117

Figure A33 Third/Fourth floor plans of Hospital (height 4.27 m), all dimensions in meters 118

Figure A34 Fifth floor plans of Hospital (height 4.27 m), all dimensions in meters 119

Figure A35 Occupancy schedules for Hospital 122

Figure A36 First floor plan of Outpatient Health Care, all dimensions in meters 127

Figure A37 Second floor plan of Outpatient Health Care, all dimensions in meters 128

Figure A38 Third floor plan of Outpatient Health Care, all dimensions in meters 129

Figure A39 Occupancy schedule for Outpatient Health Care 134

Figure A40 Floor plan of Warehouse (height 8.534 m, except for Office
    which is 4.267 m high) 135

Figure A41 Occupancy schedule for Warehouse 136

Figure A42 Floor plan of Midrise Apartment (height 3.05 m) 138

Figure A43 Occupancy schedules for Midrise Apartment 139

1. INTRODUCTION

Heating, ventilating, and air conditioning (HVAC) systems in buildings are designed to provide thermally comfortable conditions and to maintain acceptable indoor air quality (IAQ). At the same time, the operating costs of HVAC systems are often a large percentage of the total energy consumption of buildings, which constitutes 40 % of the primary energy consumed in the U.S. (DOE 2010). In order to address the need to reduce the building sector's contribution to the nation's energy consumption, a number of organizations and government agencies have set energy-related goals and are pursuing research and other activities to support achieving those goals. The American Society of Heating, Refrigerating, and Air-Conditioning Engineers (ASHRAE) has set research goals for developing standards and design guides for achieving cost-effective net-zero energy buildings (ASHRAE 2010c). The American Institute of Architects has set carbon-neutral goals for all new and major-renovated buildings by 2030 (AIA 2006). The U.S. Department of Energy (DOE) has set net-zero energy goals for both residential and commercial buildings by 2025 (DOE 2008). The DOE Building Technologies Program (BTP) supports research and development (R&D) activities to achieve these goals by improving the efficiency of buildings. One of these R&D activities is the development of the building energy simulation software EnergyPlus and its application to analyze building energy consumption and energy efficiency opportunities. Under the BTP, 16 building models were created in EnergyPlus to characterize more than 60 % of the commercial building stock in the U.S. (Deru et al. 2011). These "reference" buildings include 15 commercial buildings and one multi-family residential building. The commercial buildings include two restaurants, two health care centers, two hotels, three office buildings, two schools, three retail buildings, and a warehouse. There are three versions (or vintages) of each reference building: new, post-1980, and pre-1980 construction. The three vintages differ in insulation values, infiltration rates, lighting levels, and type of HVAC systems. The new construction models were developed to comply with the minimum requirements of ANSI/ASHRAE/IESNA Standard 90.1-2004 (ASHRAE 2004), the post-1980 models to comply with the minimum requirements of Standard 90.1-1989 (ASHRAE 1989), and the pre-1980 models to comply with requirements from previous standards and other studies of construction practices.

The reference buildings were created to assess new technologies and support the development of energy codes and standards, and therefore their definitions are focused on capturing energy performance. However, some discussions of building energy efficiency neglect potential impacts on indoor air quality (IAQ) or view acceptable IAQ as being in conflict with energy efficiency (Persily and Emmerich 2012). However, saving energy at the expense of IAQ has the potential to significantly impact the health, comfort, and productivity of building occupants. In addition, there are many approaches to building design and operation that can improve both energy efficiency and IAQ, such as heat recovery ventilation, demand control ventilation and economizer operation (Persily and Emmerich 2012). One limitation in the implementation of certain energy efficiency technologies and the consideration of their impacts on IAQ is that current energy design and analysis tools are limited in their ability to model building airflow and IAQ in a physically reasonable fashion.

A review of the airflow and IAQ analyses capabilities of five of the most widely used energy simulation software tools, including EnergyPlus, found that many of the infiltration models employed by energy simulation software are based on calculation methods developed for

1

low-rise, residential buildings (Ng and Persily 2011). These methods are not generally appropriate for other types of buildings, specifically taller buildings with mechanical ventilation systems, more airtight separations between floors, and vertical shafts. Also, these empirical infiltration models require the user to specify air leakage coefficients that are best obtained from building pressurization tests (ASTM 2010), for which only limited data are available for larger buildings (Emmerich and Persily 2011). Many energy simulation software users assume constant infiltration rates, which do not reflect known dependencies on indoor-outdoor conditions and ventilation system operation. However, airflow calculations, using existing theory and methods, are the only technically sound means of determining the airflow rates that are important for analyzing energy use and IAQ.

EnergyPlus has a so-called "Airflow Network" capability that implements multizone airflow theory, based on an earlier and limited version of AIRNET (Walton 1989) and COMIS (Feustal and Smith 2001). Airflow Network can calculate infiltration rates arising from pressure differences due to indoor-outdoor conditions and ventilation system operation. It can also model ventilation and duct systems but is limited to only one air handling system per building and only a constant volume fan. The Airflow Network capability was not incorporated into the EnergyPlus models of the reference buildings in order to simplify modeling and reduce simulation times (Deru et al. 2011).

As described in this report, models of the 16 reference buildings were created (new, post-1980, and pre-1980 versions) in the current version of CONTAM (3.0) in order to perform airflow and IAQ analyses. Together the EnergyPlus and CONTAM models allow more physically realistic analyses of the energy and IAQ impacts of envelope airtightness and airflow-related building retrofits and upgrades. The availability of the CONTAM models also supports the study of technologies and approaches that can simultaneously reduce building energy consumption while maintaining or improving IAQ.

The airflow analyses in this study included calculations of outdoor air change rates due to infiltration only, and for the combination of mechanical ventilation and infiltration. The relationships between these outdoor air change rates and weather conditions were also examined. IAQ analyses included simulations of a limited set of outdoor and indoor contaminants. Comparisons were made between airflow and energy modeling approaches, assumed and calculated infiltration rates and their impact on sensible loads, relationships between infiltration and weather, and contaminant concentrations relative to relevant standards and guidelines.

This report describes the building models and the results of the simulations of building airflow and IAQ. Section 2 describes the reference buildings and compares the CONTAM and EnergyPlus models. Section 3 describes the assumptions made in each EnergyPlus and CONTAM model in terms of airflow. Contaminants were only modeled in CONTAM, and the assumptions on which those simulations are based are described in Section 3.1. Section 4 presents the CONTAM airflow simulations for selected building models and compares the results to the airflow assumptions in the EnergyPlus models, including the impacts of these airflows on sensible loads. Section 5 presents the contaminant concentrations calculated by CONTAM. Lastly, Section 6 discusses the results of this study as well as opportunities for additional work.

2. BUILDING DESCRIPTIONS

This section provides a brief description of the reference buildings. For detailed descriptions, see Deru et al. (2011) and Appendix A, Section A2 of this report. Table 1 lists the 16 reference buildings and their floor area, number of floors, and number of zones in the EnergyPlus and CONTAM models. The number of zones is different between the two models in cases where the CONTAM models need additional zones to support more realistic airflow and IAQ analyses. For example, zones that were added to the CONTAM models include restrooms, stairwells, elevator shafts, and storage rooms (see Appendix A for more details). In some buildings, zones were also resized in order to create more realistic access between adjacent zones. For instance, in the schools, some zones were made smaller in order to create access to them from the corridor (Appendix A, Section A2.6 and A2.7). In some buildings, "multipliers" are used in the EnergyPlus models to indicate that the thermal load for one particular zone is to be applied to several other ones. This technique is employed to eliminate the need to model each individual zone in EnergyPlus. Zones with multipliers were either on the same floor or on different floors, depending on the building configuration. An example of multiplied zones on the same floor is in the Hospital (Appendix A, Section A2.13), where multiplied zones are located on the same floor since each floor has a unique layout. In the office buildings, each floor has the same layout. Thus, each zone is multiplied by the number of floors in the building. However, modeling all or at least more of the building zones is generally important for airflow and IAQ analyses. Therefore, when multiplied zones in EnergyPlus were on the same floor, they were modeled as one large zone in CONTAM. When zones with multipliers were on different floors, they were modeled as separate zones in CONTAM with leakage between them. Though zone areas and the number of zones may be different between the EnergyPlus and CONTAM models, the total building area is consistent between the two models. Further, the CONTAM models employed the occupancy and outdoor air ventilation requirements that were modeled in EnergyPlus. Details on occupancy schedules and ventilation requirements are found in Appendix A, Section A2.

3. MODELING APPROACH

EnergyPlus simulations were performed for 16 U.S. cities, which represent eight climates zones and cover 78 % of the U.S. population (Deru et al. 2011). CONTAM simulations were only performed using weather data from Chicago, IL, since there are a relatively high percentage of buildings in the U.S. in this climate zone (Deru et al. 2011). Also, the system airflows calculated by EnergyPlus for Chicago were in the mid-range of HVAC airflow rates calculated for all 16 cities. Typical meteorological year, version 2 (TMY2) weather data were obtained from DOE (DOE 2011), which contain outdoor temperature, outdoor humidity, and wind direction and speed. Two weather files were created for use with the CONTAM models – one for the schools (Primary and Secondary) and the other for the remaining buildings. The difference between the two weather files was the use of a "special day" to indicate the use of a "summer" occupancy schedule for use in transient CONTAM simulations.

Section 3.1 presents the airflow and contaminant inputs in the CONTAM models. Section 1.1 presents the airflow inputs in the EnergyPlus models, as no contaminants were included in the EnergyPlus models.

Table 1 Summary of reference buildings

Building	Floor area (m^2)	No. of floors	No. of EnergyPlus zones	No. of CONTAM zones
Restaurants				
Full service	511	1	2	3[3]
Quick service	232	1	2	3[3]
Health care centers				
Hospital	22422	6[1]	55[1]	64[1,3,4]
Outpatient	3804	3	118[2]	118[2]
Hotels				
Small	4013	4	67[2]	67[2]
Large	11345	6	43	49[3,4]
Offices				
Small	511	1	5	6[3]
Medium	4982	3	18	23[3,4]
Large	46320	13[1]	73[1]	87[1,3,4]
Schools				
Primary	6871	1	25	25
Secondary	19592	2	46	46
Retail				
Stand-alone	2294	1	5	6[3]
Strip mall	2090	1	10	30[3,5]
Supermarket	4181	1	6	6
Warehouse	4835	1	3	4[3]
Midrise apartment	3135	4	36	38[4]

1. Includes a basement.
2. Includes a stairwell and elevator shaft.
3. Includes restroom(s) not in the EnergyPlus models.
4. Includes stairwell(s) and elevator shaft(s) not in the EnergyPlus models.
5. Includes storage rooms not in the EnergyPlus models.

3.1. CONTAM model inputs

Building exterior envelope leakage was modeled in CONTAM using an effective leakage area (A_L) of 5.27 cm^2/m^2 at a reference pressure difference (ΔP_r) of 4 Pa, a discharge coefficient (C_D) of 1.0, and a pressure exponent (n) of 0.65 for all three vintages of the reference buildings. This leakage area value was based on consideration of airtightness data in U.S. commercial buildings (Emmerich and Persily 2011), which does not support the use of different values for the different vintages. This envelope leakage was applied to all above-grade exterior walls, ceilings, roofs, and floors. Basement walls were modeled with half of the leakage specified for above grade walls, and slab floors were modeled with no leakage. The infiltration airflow through these leaks is calculated by CONTAM using a power-law relationship:

$$Q = \frac{C_D A_L}{10000} \sqrt{\frac{2}{\rho}} \left(\Delta P_r\right)^{0.5-n} \Delta P^n \qquad (1)$$

where the indoor-outdoor pressure difference (ΔP) is calculated by CONTAM based on wind and stack effects and ventilation equipment operation, as well as interior zone pressure relationships. CONTAM also calculates the air density (ρ) based on the temperature of the air entering the leakage site. Detailed discussion on how CONTAM calculates these values is found in Walton and Dols (2005). To capture the stack effect more accurately, exterior wall leakage was divided into three portions on each wall, representing the lower third, middle third, and upper third of each wall. Wind effects are calculated using a wind pressure profile, which describes the wind pressure coefficients (C_P) as a function of wind directions (θ). Figure 1 is the wind pressure profile that was used (Swami and Chandra 1987). A wind speed modifier of 0.36, which corresponds to "suburban" terrain (Walton and Dols 2005), was applied to all exterior leakage paths.

Figure 1 Wind pressure profile simulated in CONTAM for exterior walls

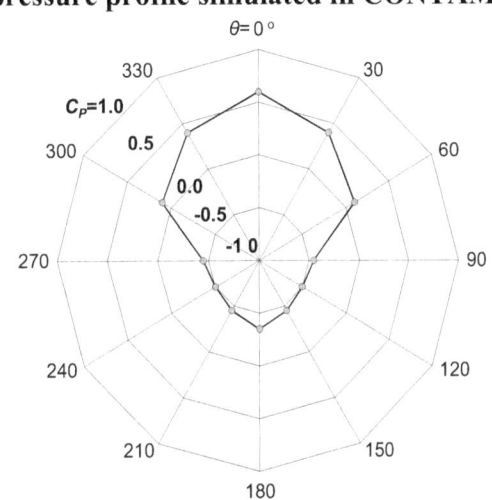

For openings on roofs, C_P was -0.5 for all wind directions. This was an average value for roofs with less than a 15 degree slope shown in the ASHRAE Handbook of Fundamentals, Chapter 24 (ASHRAE 2009). For buildings with attics, leakage from the roof was modeled with venting equal to 1/150 of the floor area (Lstiburek 2006).

The effective leakage area of partitions between floors and between zones used the same value as the exterior wall leakage (5.27 cm^2/m^2 at 4 Pa). The connections between zones that would not have a physical partition, such as within an open office or retail space, were modeled as large openings with discharge coefficient C_D=0.6 and n=0.5. The size of these openings ranged from 50 % to 75 % of the wall area between zones. Transfer grilles, ranging from 0.186 m^2 to 0.372 m^2, and door undercuts of 0.025 m^2, were modeled between restrooms and adjacent zones.

The minimum amount of outdoor ventilation air for each zone (or HVAC system) was specified in EnergyPlus using ASHRAE 62-1999 (ASHRAE 1999). Depending on the thermal load calculated at each time step and the indoor-outdoor conditions, EnergyPlus varied the amount of outdoor ventilation. However, the minimum amount of outdoor ventilation air for each zone (or HVAC system) was modeled in CONTAM with no economizer cycle to simplify the modeling inputs. Details on the supply, return, and outdoor ventilation rates modeled in CONTAM can be found in Appendix A.

The Energy Plus models have a variety of system types, ranging from through-the-wall packaged single zone systems to variable-air volume (VAV) air handling units. All air handling units serving multiple zones were modeled using the "simple air handling system" model in CONTAM. The simple air handling unit components were modeled as follows:

- The volume of supply ductwork was specified as 1 % of the building volume served by the system.
- The volume of return ductwork was specified as 0.5 % of the building volume served by the system (if no return air plenum is present) and 0.25 % of the building volume served (if a return air plenum is present).
- Supply and return diffusers were located in the zones to provide the design (or maximum) airflow rate calculated by EnergyPlus. For VAV systems included in the EnergyPlus models, only the maximum airflow rates calculated by EnergyPlus were modeled in CONTAM.
- Outdoor ventilation rates were specified at the air handling unit to provide the minimum ventilation rate specified in EnergyPlus.

Where single-zone packaged units were modeled in EnergyPlus, a constant mass flow element was modeled in CONTAM to add or remove the appropriate amount of outdoor air from the zone. Constant mass flow elements were also used to represent restroom and kitchen exhaust fans. Where these fans were included in the EnergyPlus models, those exhaust flow rates were used in the CONTAM models. Restrooms were not included in the EnergyPlus models of the Full Service Restaurant, Medium Office, and Stand-Alone Retail. In these buildings, restrooms were added to the CONTAM models and the flow rates were based on ASHRAE Standard 62.1-2010 (ASHRAE 2010a).

Detailed duct models were not modeled in either CONTAM or EnergyPlus. However, for interior restrooms (or other zones with exhaust fans), a single length of duct was used in CONTAM to connect the restroom exhaust to the outdoors. The segment was specified with a constant mass flow rate so that the physical characteristics of the ducts did not influence the resulting airflow rate.

The common design goal of pressurizing commercial buildings was accounted for in the CONTAM models by returning 90 % of the supply airflow rate. When the outdoor air quantity to a zone was less than 10 % of the supply, the return airflow rate was equal to the supply minus the outdoor airflow rate. For buildings with large exhaust fans, i.e., the two restaurants, the total outdoor air intake was approximately equal to the total exhaust.

Contaminant simulations were performed for four contaminants in CONTAM: carbon dioxide (CO_2), ozone, particulates less than 2.5 µm in diameter (PM 2.5), and a generic volatile organic compound (VOC). Table 2 summarizes the properties of these contaminants. Outdoor concentrations of ozone and PM 2.5 were downloaded from the U.S. Environmental Protection Agency (EPA) Air Quality Standard (AQS) database (EPA 2011a). Table 3 lists the minimum, maximum, mean, and standard deviation of the outdoor concentration of ozone and PM 2.5 for Chicago, IL. The outdoor concentration of CO_2 was assumed to be 648 mg/m^3 and that of VOC was assumed to be zero for the CONTAM simulations.

Table 2 Properties of contaminants simulated in CONTAM

Contaminant	Molecular Weight	Mean Diameter (µm)	Effective Density (kg/m^3)
CO_2	44	N/A	N/A
Ozone	48	N/A	N/A
PM 2.5	N/A	0.3[1]	1,000[2]
VOC	92[3]	N/A	N/A

1. Based on measurements in urban areas (Riley et al. 2002).
2. Equal to density of water (Chen and Zhao 2011; Riley et al. 2002).
3. Based on toluene as a representative VOC.

Table 3 Summary of outdoor contaminant concentrations for Chicago

Outdoor contaminant	Daily average contaminant concentrations				Daily peak contaminant concentrations			
	Mean	Min.	Max.	StdDev	Mean	Min.	Max.	StdDev
Ozone, µg/m^3	47	6	106	21	80	12	155	29
PM 2.5, µg/m^3	18	1	57	10	30	4	94	14

Indoor contaminant sources included occupant-generated CO_2 and VOCs from materials and activities. A CO_2 source was defined in all occupied zones, with an assumed generation rate of 0.3 L/min per person (ASHRAE 2010a). The maximum number of occupants specified in the EnergyPlus models was used in the CONTAM models. The CO_2 source strength in the CONTAM models varied with occupancy based on schedules in the EnergyPlus models. Detailed occupancy schedules for each building are found in Appendix A. An area-based VOC source was defined in all occupied building zones. In occupied zones, a 0.5 mg/m^2•h source was included during system-on hours and reduced by 50 % during system-off hours (Persily et al. 2003). Zones that were always unoccupied had no VOC source. Deposition rates of 0.5 h^{-1} for PM 2.5 (Allen et al. 2003; Howard-Reed et al. 2003; Riley et al. 2002) and 4.0 h^{-1} for ozone (Kunkel et al. 2010; Nazaroff et al. 1993; Weschler 2000; Weschler et al. 1989) were included in every zone. No indoor sources were included for ozone or PM 2.5.

A constant efficiency filter was placed in both the outdoor and recirculation air streams of all HVAC systems in the CONTAM models (including supply air delivered by the packaged terminal air conditioning (PTAC) units in the Small Hotel) to represent a filter placed in the mixed air stream. The filter removed ozone at 5 % efficiency (Bekö et al. 2006) and removed PM 2.5 at 25 % efficiency, corresponding to filters with a Minimum Efficiency Reporting Value (MERV) of 6 as required in ASHRAE Standard 62.1-2010 (ASHRAE 2010a; Kowalski and Bahnfleth 2002). A penetration factor of one was assumed for both ozone (Liu and Nazaroff 2001; Weschler et al. 1989) and PM 2.5 (Allen et al. 2003; Thornburg et al. 2001; Tian et al. 2009), i.e., there was no removal of these contaminants in the exterior leakage paths.

3.2. EnergyPlus model inputs

A simplified approach for modeling infiltration was used in the EnergyPlus models of the reference buildings in order to simplify the assumptions needed and to reduce simulation times (Deru et al. 2011). For the EnergyPlus models of the "new" buildings, building envelope leakage was assumed to be 1.18 cm^2/m^2 at a constant indoor-outdoor pressure of 4 Pa, based on a *proposed* addendum to ASHRAE 90.1-2004 for an air barrier requirement (Deru et al. 2011). Using Equation (1), this building envelope leakage is equivalent to an airflow rate at 4 Pa of 0.000 302 $m^3/s\cdot m^2$ of exterior surface area, which is input into the EnergyPlus models as the building infiltration rate.

It should be noted that there is now an air barrier requirement in ASHRAE 90.1-2010 (ASHRAE 2010b), but it does not contain a whole building airtightness requirement but rather only material and assembly tightness requirements. Note also that the value of 1.18 cm^2/m^2 used in the EnergyPlus models is not necessarily consistent with expectations for buildings of this vintage based on the existing airtightness data. However, the value of 5.27 cm^2/m^2 used in the CONTAM models is supported by consideration of these data (Emmerich and Persily 2011). Besides the building leakage value, perhaps a more important difference between the CONTAM and EnergyPlus models is that the indoor-outdoor pressure difference across the exterior envelope is actually calculated in CONTAM rather than assumed to be a constant 4 Pa as in the EnergyPlus models. Assuming a constant pressure difference does not reflect known dependencies of infiltration on indoor-outdoor pressure differences. Comparisons between the assumed infiltration rates in EnergyPlus and those calculated by CONTAM are discussed in Section 4.

Infiltration was scheduled at 100 % of the input value when the ventilation system was scheduled to be off and reduced to 25 % or 50 % when the ventilation system was scheduled to be on. The exceptions to this approach were the Full Service Restaurant, Primary School, and Stand-Alone Retail. In the Full Service Restaurant, the HVAC system was scheduled to be on from 5 a.m. to 1 a.m. However, from 12 a.m. to 1 a.m., the EnergyPlus models had infiltration scheduled at 100 % rather than a reduced value. In the Primary School, the HVAC system was scheduled to be on from 6 a.m. to 9 p.m. However, from 6 a.m. to 7 a.m., infiltration was scheduled at 100 % rather than a reduced value. On weekends and holidays, the Primary School HVAC system was scheduled to always be off. On these days, infiltration was scheduled at 50 % from 7 a.m. to 9 p.m. rather than the full value. In the Stand-Alone Retail, the HVAC system was scheduled to be on from 8 a.m. to 7 p.m. on Sundays and holidays. However, from 5 p.m. to 7 p.m., infiltration was scheduled at 100 % rather than a reduced value.

It should also be noted that between zones for which no physical partition would actually exist, a physical wall (two layers of ½" gypsum) was modeled between the zones in EnergyPlus. The walls were modeled in this way to produce temperature differences between the zones, but no airflow between these zones was modeled in EnergyPlus. In other selected zones, a simplified approach for interzone airflow was taken. For instance, in buildings with kitchens or dining areas, the modeling convention was to specify a constant airflow rate from the dining area to the kitchen area. An exhaust fan was then modeled in the dining area at a removal rate equal to this airflow rate in order to balance flows in the zones. There was no airflow between building floors in the EnergyPlus models.

In most zones of the buildings, the EnergyPlus HVAC systems were modeled with the supply airflow rate equal to the return airflow rate. In some zones, such as dining and kitchen zones, there was an excess of exhaust, so that the return was reduced such that the sum of the return and exhaust equalled the supply. The effect of infiltration (either from outside or adjacent zones) on thermal loads is considered in EnergyPlus, but infiltration is not part of the mass balance of air into or out of the zones.

Economizer operation increased the ventilation rate above the minimum requirement when the following conditions were met: (1) outdoor temperature less than 28 C; (2) return temperature to the HVAC system greater than the mixed air temperature after the outdoor air mixing box; and (3) return temperature to the HVAC system greater than the outdoor temperature. Table 4 shows the buildings for which at least one economizer was modeled in EnergyPlus in Chicago, IL. Details on the economizer in EnergyPlus for each building type, vintage, and city can be found on the DOE website (DOE 2011).

Table 4 Buildings for which economizer modeled for at least one HVAC system (Chicago)

Building	Economizer in new building?	Economizer in post-1980 building?	Economizer in pre-1980 building?
Restaurants			
Full service	Y	Y	Y
Quick service	N	N	N
Health care centers			
Hospital	Y	Y	Y
Outpatient	Y	Y	Y
Hotels			
Small	N	N	N
Large	Y	Y	Y
Offices			
Small	N	N	N
Medium	Y	Y	Y
Large	Y	Y	Y
Schools			
Primary	Y	Y	Y
Secondary	Y	Y	Y
Retail			
Stand-alone	Y	Y	Y
Strip mall	N	Y	Y
Supermarket	Y	Y	Y
Warehouse	Y	Y	Y
Midrise apartment	N	N	N

4. AIRFLOW SIMULATION RESULTS

Airflow simulations were performed for six building models representing each type of occupancy covered by the 15 commercial reference buildings, which exclude the Midrise Apartment building. The buildings simulated were: Full Service Restaurant, Hospital, Medium Office, Primary School, Small Hotel, and Stand-Alone Retail. Annual simulations for the "new" buildings were performed in both CONTAM and EnergyPlus. EnergyPlus results are presented using two different infiltration rates. EnergyPlus results using the building envelope leakage value assumed in Deru et al. (2011) (1.18 cm^2/m^2 at 4 Pa) are referred to as "EnergyPlus (tight)." EnergyPlus results using the building envelope leakage value assumed in the CONTAM models (5.27 cm^2/m^2 at 4 Pa) are referred to as "EnergyPlus (CONTAM-equivalent)."

The timestep for the CONTAM simulations was 1 hour, since the TMY2 weather data was hourly. The timesteps for the EnergyPlus simulations were 10 minutes or 15 minutes depending on the building. A shorter time step was used in EnergyPlus in order to better capture the effects of heat transfer and equipment operation.

Section 4.1 presents the distribution of outdoor air change rates for the six simulated buildings, and compares the rates calculated by CONTAM and EnergyPlus. Section 4.2 discusses the outdoor air change rates calculated by CONTAM and EnergyPlus as a function of weather conditions. Lastly, Section 4.3 presents the impact of infiltration on the sensible heating and cooling loads for the six simulated buildings.

4.1. Outdoor air change rates

Outdoor air change rates were calculated as the total flow of outdoor air into the building (including both air leakage through the exterior envelope and outdoor air intake via the mechanical ventilation system) divided by the building volume. Attics were not included in the building volume. It should be noted that the system is scheduled to be on when the majority of the occupants are present in the building. The systems are generally off when the buildings are not occupied, and in some cases, when the buildings are minimally occupied. In the Hospital and Small Hotel, the HVAC system is always on.

Table 5 lists the number of hours under each system/fan condition, as well as the corresponding minimum, maximum, mean, and standard deviation of the outdoor air change rates calculated by CONTAM and EnergyPlus. The EnergyPlus values include the two assumed infiltration rates, tight and CONTAM-equivalent. Values are listed for both system-on and system-off conditions. "Economizer air change rates," in the right side of Table 5, correspond to times when the HVAC system was scheduled to be off but the controller for the economizer turned the fan on because of suitable indoor-outdoor conditions. Economizers also operated during system-on hours, and those air change rates are included in the system-on values in Table 5. Economizers were modeled in all EnergyPlus models of the buildings except the Small Hotel. As noted earlier, no economizers were modeled in CONTAM. Table 5 also lists the outdoor air change rate due to infiltration only when the system is on. Figure 2 to Figure 7 show the frequency distribution of outdoor air change rates for each of the six buildings for an entire year. Each figure presents the air change rates for the various system conditions, on and off, CONTAM and Energy Plus, and economizer operation.

Outdoor air change rates with system on

Table 5 shows that the Full Service Restaurant has the highest system-on outdoor air change rates, with the following mean values: CONTAM 4.83 h^{-1}, EnergyPlus (tight) 6.57 h^{-1}, and EnergyPlus (CONTAM-equiv.) 6.61 h^{-1}. The high outdoor air change rates in the Full Service Restaurant are due to (1) the amount of ventilation air needed as makeup air for the large kitchen exhaust rates, and (2) high occupant density combined with the fact that the outside air supplied per person is the third highest among the simulated buildings (Table 7 in Section 4.3). The next highest system-on outdoor air change rates occur in the Primary School. It has a large kitchen exhaust fan and the second highest occupant density of the simulated buildings. The Medium Office and Hospital have the lowest system-on outdoor air change rates as calculated using CONTAM. Low outdoor air change rates in these two buildings are due to low occupant density, despite their having the second highest and highest, respectively, outside air supplied per person (Table 7).

As expected, the system-on outdoor air change rates calculated using the EnergyPlus (CONTAM-equiv.) models are *higher* than those calculated using the EnergyPlus (tight) models since the infiltration rate input to the EnergyPlus (CONTAM-equiv.) models was about four times the rate input to the EnergyPlus (tight) models. When comparing the system-on outdoor air change rates calculated using CONTAM and the EnergyPlus models (tight and CONTAM-equiv.), the results are different for each building.

For the Full Service Restaurant, Hospital, and Medium Office, the mean system-on outdoor air change rates calculated using CONTAM were approximately 30 % to 40 % lower than the EnergyPlus (tight and CONTAM-equiv.) results. For the Stand-Alone Retail, the CONTAM results were only 4 % lower than the EnergyPlus (tight) results, but 40 % lower than the EnergyPlus (CONTAM-equiv.) results. For the Primary School, the CONTAM results were 20 % to 40 % higher than the EnergyPlus, tight and CONTAM-equiv., results respectively. For the Small Hotel, the CONTAM result was 60 % higher than the EnergyPlus (tight) result, but 10 % lower than the EnergyPlus (CONTAM-equiv.) result.

Table 5 Summary of calculated outdoor air change rates

Full Service Restaurant

Full Service Restaurant	System-on outdoor air change rates, h^{-1}					System-on infiltration rates, h^{-1}				System-off outdoor air change rates, h^{-1}					Economizer air change rates, h^{-1}				
	Hours	Mean	Min.	Max.	StdDev	Mean	Min.	Max.	StdDev	Hours	Mean	Min.	Max.	StdDev	Hours	Mean	Min.	Max.	StdDev
CONTAM	7300	4.83	4.31	6.16	0.27	0.53	0.01	1.86	0.27	1460	0.50	0.00	1.87	0.25					
EnergyPlus (T*)	7300	6.57	6.27	7.28	0.20	0.10	0.09	0.19	0.02	1460	0.19	0.19	0.19	0.00					
EnergyPlus (CE)	7300	6.94	6.61	8.10	0.25	0.46	0.43	0.88	0.09	1435	0.87	0.87	0.88	0.00					
															25	3.40	1.91	4.08	0.68

Hospital

Hospital	System-on outdoor air change rates, h^{-1}					System-on infiltration rates, h^{-1}				System-off outdoor air change rates, h^{-1}				
	Hours	Mean	Min.	Max.	StdDev	Mean	Min.	Max.	StdDev	Hours	Mean	Min.	Max.	StdDev
CONTAM	8760	0.91	0.85	1.23	0.05	0.05	0.00	0.37	0.05	0	NA	NA	NA	NA
EnergyPlus (T)	8760	1.28	0.95	2.05	0.32	0.01	0.01	0.01	0.00					
EnergyPlus (CE)	8760	1.32	0.99	2.20	0.32	0.04	0.04	0.04	0.00					

Medium Office

Medium Office	System-on outdoor air change rates, h^{-1}					System-on infiltration rates, h^{-1}				System-off outdoor air change rates, h^{-1}					Economizer air change rates, h^{-1}				
	Hours	Mean	Min.	Max.	StdDev	Mean	Min.	Max.	StdDev	Hours	Mean	Min.	Max.	StdDev	Hours	Mean	Min.	Max.	StdDev
CONTAM	4644	0.68	0.09	1.35	0.16	0.12	0.00	0.75	0.11	4116	0.28	0.00	0.86	0.13					
EnergyPlus (T)	4644	0.93	0.05	1.93	0.31	0.05	0.05	0.05	0.00	3564	0.20	0.20	0.21	0.00					
															552	0.34	0.21	1.09	0.20
EnergyPlus (CE)	4644	1.09	0.31	2.07	0.30	0.23	0.22	0.23	0.00	2644	0.91	0.90	0.92	0.00	1472	1.02	0.91	1.75	0.07

* T corresponds to EnergyPlus (tight) case, and CE corresponds to EnergyPlus (CONTAM-equiv.) case.

Note: No economizer was modeled in CONTAM. Only for the Medium Office and Primary School (EnergyPlus (tight) models) were indoor-outdoor conditions suitable for economizer operation when the HVAC system was scheduled to be off. In addition, there was economizer operation for the Full Service Restaurant (EnergyPlus (CONTAM-equiv.) model). The calculated air change rates of $0.00\ h^{-1}$ during system-off hours using CONTAM correspond to very small indoor-outdoor temperature differences and/or wind speeds. The values are not exactly zero but are less than $0.005\ h^{-1}$.

(Table 5 continued)

Primary School

	System-on outdoor air change rates, h⁻¹					System-on infiltration rates, h⁻¹				System-off outdoor air change rates, h⁻¹				
	Hours	Mean	Min.	Max.	StdDev	Mean	Min.	Max.	StdDev	Hours	Mean	Min.	Max.	StdDev
CONTAM	3780	1.88	1.65	2.74	0.14	0.32	0.09	1.17	0.14	4980	0.29	0.01	0.97	0.14
EnergyPlus (T)	3780	1.36	1.18	2.40	0.22	0.05	0.05	0.10	0.01	4032	0.08	0.05	0.10	0.02

Economizer air change rates, h⁻¹

	Hours	Mean	Min.	Max.	StdDev
	948	0.22	0.06	0.77	0.09

	Hours	Mean	Min.	Max.	StdDev	Mean	Min.	Max.	StdDev	Hours	Mean	Min.	Max.	StdDev
EnergyPlus (CE)	3780	1.55	1.40	2.57	0.19	0.24	0.22	0.45	0.06	3034	0.38	0.22	0.45	0.11

Economizer air change rates, h⁻¹

	Hours	Mean	Min.	Max.	StdDev
	1946	0.57	0.24	0.92	0.13

Small Hotel

	System-on outdoor air change rates, h⁻¹					System-on infiltration rates, h⁻¹				System-off outdoor air change rates, h⁻¹				
	Hours	Mean	Min.	Max.	StdDev	Mean	Min.	Max.	StdDev	Hours	Mean	Min.	Max.	StdDev
CONTAM	8760	1.04	0.78	1.97	0.15	0.26	0.00	1.19	0.15	0	NA	NA	NA	NA
EnergyPlus (T)	8760	0.64	0.58	0.81	0.04	0.14	0.14	0.14	0.00					
EnergyPlus (CE)	8760	1.13	1.08	1.31	0.04	0.63	0.62	0.64	0.00					

Stand-Alone Retail

	System-on outdoor air change rates, h⁻¹					System-on infiltration rates, h⁻¹				System-off outdoor air change rates, h⁻¹				
	Hours	Mean	Min.	Max.	StdDev	Mean	Min.	Max.	StdDev	Hours	Mean	Min.	Max.	StdDev
CONTAM	5278	1.03	0.80	1.61	0.14	0.23	0.00	0.82	0.14	3482	0.26	0.00	0.88	0.13
EnergyPlus (T)	5278	1.07	0.93	1.73	0.21	0.14	0.14	0.27	0.02	3482	0.27	0.26	0.27	0.00
EnergyPlus (CE)	5278	1.70	1.40	3.77	0.53	0.63	0.60	1.22	0.09	3482	1.21	1.20	1.23	0.00

Table 5 shows that the mean system-on infiltration rates calculated using CONTAM were about two to six times higher than the assumed inputs in the EnergyPlus (tight) models. For large temperature differences and wind speeds, the corresponding system-on infiltration rates calculated using CONTAM were as much as nine times higher than the EnergyPlus (tight) inputs. For the Full Service Restaurant, Hospital, and Primary School, the mean system-on infiltration rates calculated using CONTAM were also higher than the assumed inputs in the EnergyPlus (CONTAM-equiv.) models. However, rather than roughly five times higher as in the case of the EnergyPlus (tight) inputs, the CONTAM results were only 20 % to 30 % higher than the EnergyPlus (CONTAM-equiv.) inputs. For the Medium Office, Small Hotel, and Stand-Alone Retail, the mean system-on infiltration rates calculated using CONTAM were 50 % to 60 % lower than the EnergyPlus (CONTAM-equiv.) inputs.

Because the infiltration rates in EnergyPlus were scheduled as constant values, even when the system was on, the standard deviations of system-on infiltration rates in Table 5 are zero for most of the buildings. The standard deviations of system-on infiltrations are greater than zero for the Full Service Restaurant, Primary School, and Stand-Alone Retail because the system-on infiltration rate was scheduled using two different values (Section 1.1). Even in cases where the mean outdoor air change rates are similar, the outdoor air change rates calculated using EnergyPlus do not reflect the dependency of infiltration on weather, while the CONTAM results do. The impact of the differences in infiltration rates on energy consumption are discussed in Section 4.3.

Outdoor air change rates with system off
Table 5 shows that the *mean* system-off outdoor air change rates calculated using the EnergyPlus (CONTAM-equiv.) models were higher than those calculated using the EnergyPlus (tight) models. This is expected since the infiltration rate input in the EnergyPlus (CONTAM-equiv.) models was about four times the rate input in the EnergyPlus (tight) models.

The infiltration rates calculated by CONTAM include weather effects, unlike EnergyPlus for which infiltration rates were scheduled as constant values. Thus, the standard deviations in Table 5 of the system-off outdoor air change rates calculated using CONTAM are greater than zero, while those calculated using EnergyPlus are zero. The only exception is the Primary School, where the EnergyPlus standard deviation is greater than zero because the system-off infiltration rate was scheduled using two different values. Table 5 also shows that only for the EnergyPlus (tight) models of the Medium Office and Primary School were conditions suitable for economizer operation when the HVAC system was scheduled to be off. When the infiltration rate was increased in the EnergyPlus (CONTAM-equiv.) models, the economizer was also on for the Full Service Restaurant.

Table 5 shows that the mean system-off infiltration rates calculated using CONTAM were about 30 % to three times higher than the assumed inputs in the EnergyPlus (tight) models. An exception is seen in the Stand Alone-Retail building, where the mean values for CONTAM and EnergyPlus (tight) are very similar. For large temperature differences and wind speeds, the system-on infiltration rates calculated using CONTAM were as much as five times higher than the EnergyPlus (tight) inputs. For all buildings, the mean system-off infiltration rates calculated using CONTAM were 20 % to 80 % lower than the assumed inputs in the EnergyPlus (CONTAM-equiv.) models, independent of weather condition.

Regardless of whether the *mean* system-off infiltration rates calculated using CONTAM were lower or higher than the EnergyPlus inputs, it is important to note that the infiltration rates calculated by CONTAM take weather effects into account whereas the EnergyPlus results do not. This is clearly seen by comparing the standard deviations in Table 5 as well as the distribution of system-off outdoor air change rates in Figure 2 to Figure 7 (excluding the Hospital and Small Hotel since they have 24-hour HVAC operating schedules). The solid red bars in plots (a) of Figure 2 to Figure 7 show the variability in system-off outdoor air change rates as calculated using CONTAM. In contrast, the solid red bars in plots (b) of Figure 2 to Figure 7 show that the system-off outdoor air change rates calculated using EnergyPlus are constant, except for buildings that use two different scheduled values.

For the two EnergyPlus (tight) models in which economizer operation was suitable when the HVAC system was scheduled to be off, e.g., Medium Office and Primary School, the mean economizer outdoor air change rates calculated using EnergyPlus (tight) were larger than the system-off outdoor air change rates. Referring to Table 5, the number of system-off economizer hours for the Medium Office and Primary School was about 15 % and 25 % of their respective number of system-off hours. For the Full Service Restaurant, the increase in the assumed infiltration rate in the EnergyPlus (CONTAM-equiv.) models resulted in 25 hours of economizer operation, which was 2 % of the system-off hours. For the Medium Office and Primary School, the increase in the assumed infiltration rate in the EnergyPlus (CONTAM-equiv.) models resulted in two to three times *more* hours of economizer operation than in the EnergyPlus (tight) models.

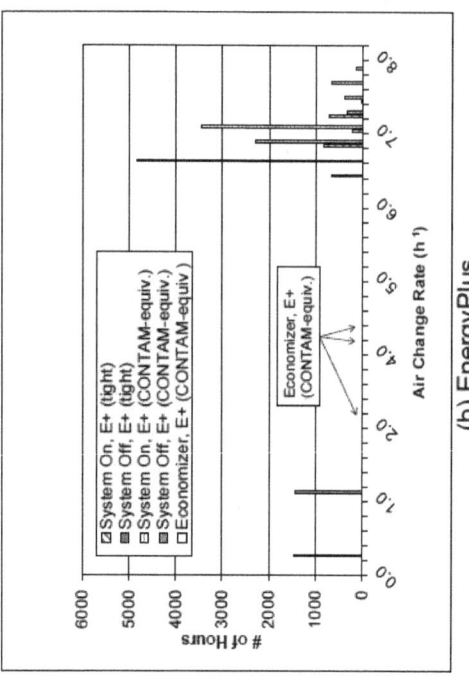

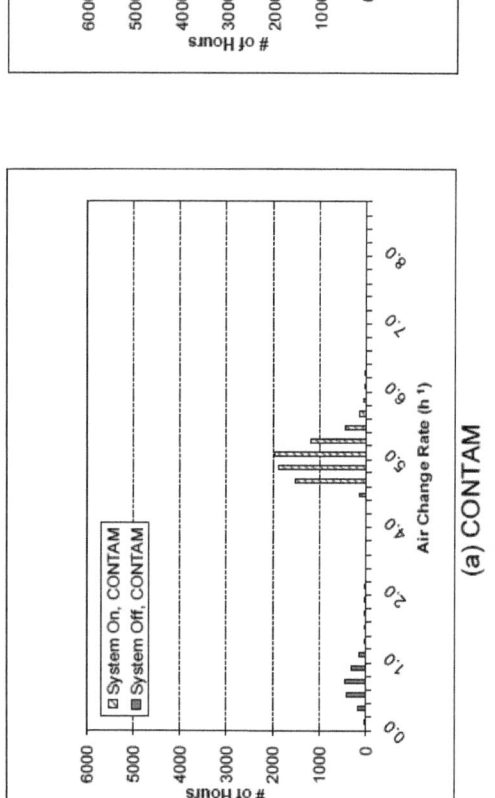

(b) EnergyPlus

(a) CONTAM

Figure 2 Frequency distribution of simulated outdoor air change rates for Full Service Restaurant

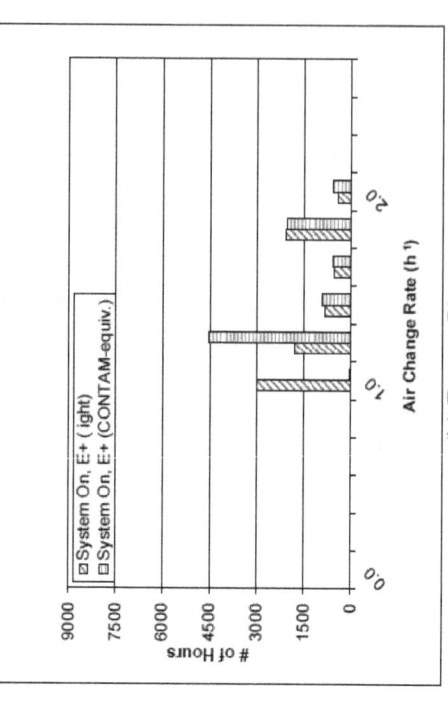

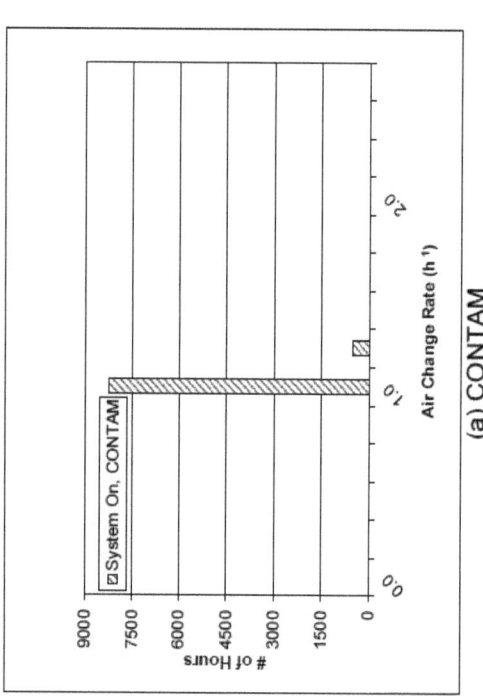

(b) EnergyPlus

(a) CONTAM

Figure 3 Frequency distribution of simulated outdoor air change rates for Hospital

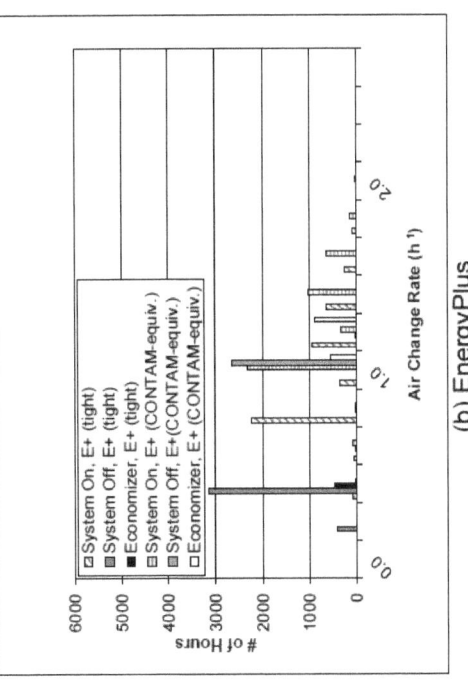

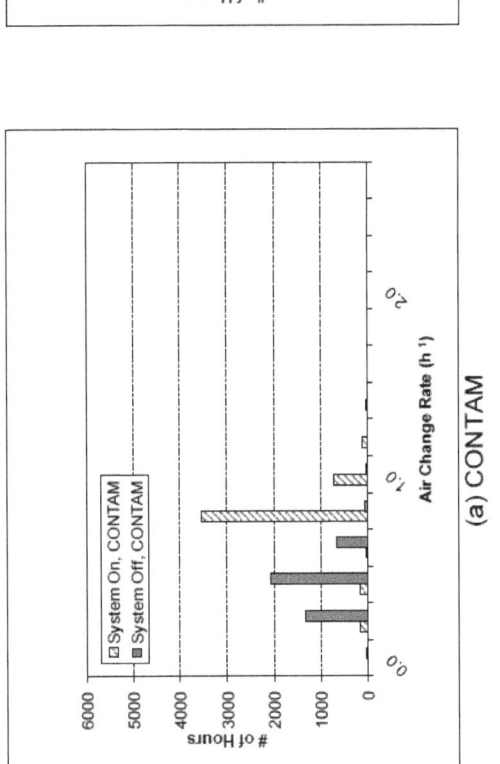

(a) CONTAM

(b) EnergyPlus

Figure 4 Frequency distribution of simulated outdoor air change rates for Medium Office

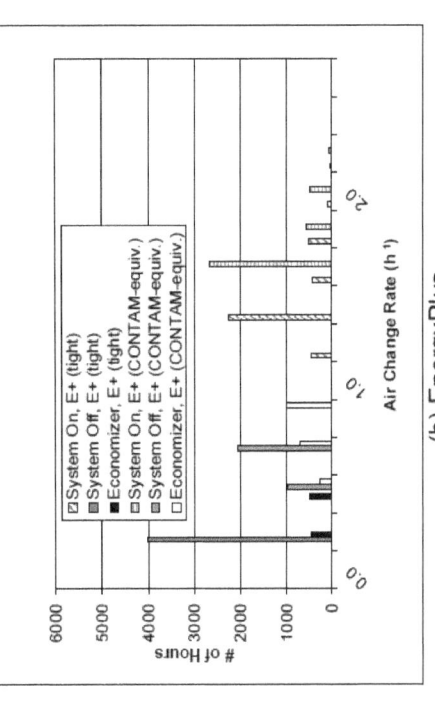

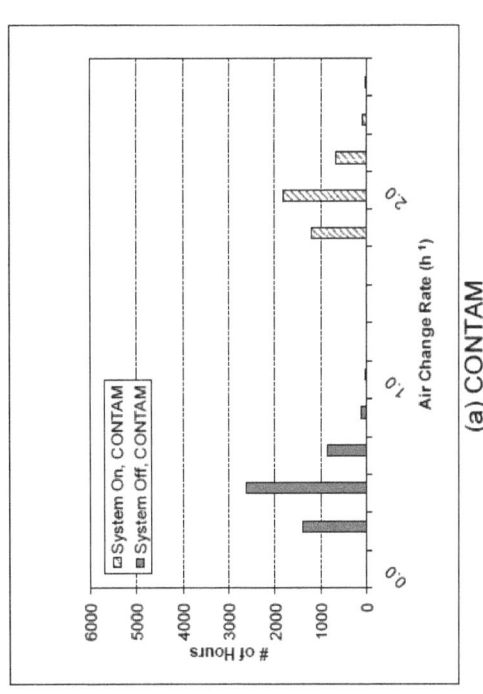

(a) CONTAM

(b) EnergyPlus

Figure 5 Frequency distribution of simulated outdoor air change rates for Primary School

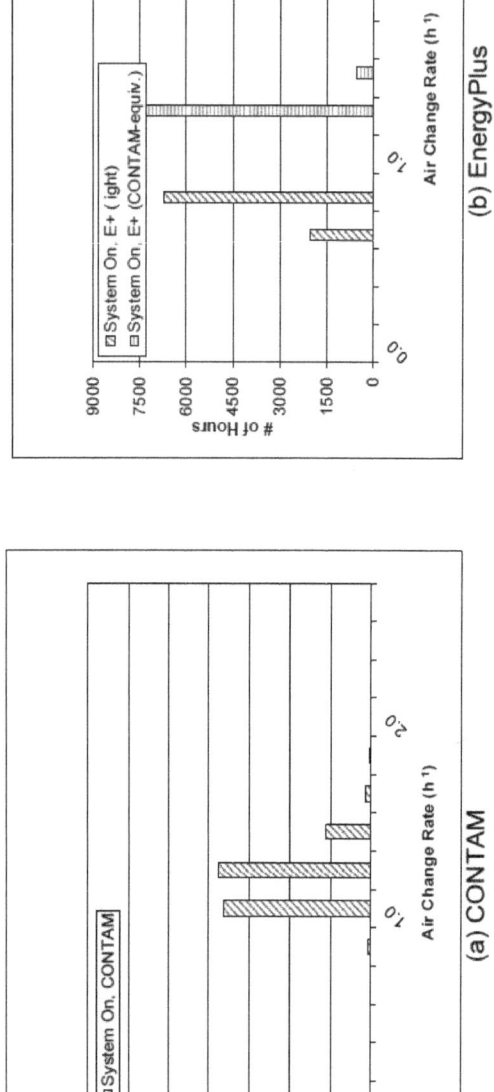

(a) CONTAM

(b) EnergyPlus

Figure 6 Frequency distribution of simulated outdoor air change rates for Small Hotel

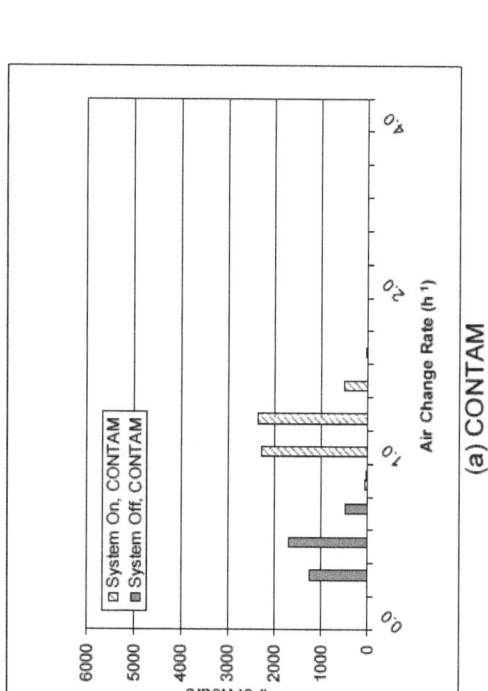

(a) CONTAM

(b) EnergyPlus

Figure 7 Frequency distribution of simulated outdoor air change rates for Stand-Alone Retail

4.2. Outdoor air change rates vs. weather conditions

Outdoor air change rates are plotted against indoor-outdoor temperature difference, ΔT (Figure 8 to Figure 13) and wind speed, W_s (Figure 14 to Figure 19). The plots of outdoor air change rate versus indoor-outdoor temperature difference only include air change rates for low wind speeds, i.e., less than 2 m/s. The plots of outdoor air change rate versus wind speed are shown for indoor-outdoor temperature differences with absolute values less than 10 $^{\circ}$C. Limiting the plots to low temperature differences and wind speeds makes the effects of ΔT and W_s rate easier to see. Results are plotted for system-on, system-off, and economizer hours (where applicable).

In EnergyPlus, cooling and heating setpoints were specified for system-on and system-off hours. Thus, the indoor temperature was not constant. When plotting the outdoor air change rates against the indoor-outdoor temperature difference, the "indoor" temperature for the EnergyPlus simulations were the average of the return air temperatures in the building HVAC systems. In CONTAM, a constant indoor temperature of 20 $^{\circ}$C was assumed.

As noted earlier, economizers were included in all of the EnergyPlus models except the Small Hotel (Section 1.1). However, only for the Medium Office and Primary School did the economizer operated when the HVAC system was scheduled to be off for the EnergyPlus (tight) models. For the EnergyPlus (CONTAM-equiv.) models, economizers operated in the Full Service Restaurant, Medium Office, and Primary School. No economizers were modeled in CONTAM.

Outdoor air change rates vs. weather for system-on hours

Plots (a) of Figure 8 to Figure 13 show that generally there is a linear relationship between system-on outdoor air change rates calculated using CONTAM and ΔT, with the dependence being symmetrical about ΔT=0. However, for the Hospital, outdoor air change rates calculated using CONTAM are not significantly affected by temperature difference as seen in Figure 9a. The Hospital HVAC system is scheduled to always be on, and the mechanical airflows dominate the envelope infiltration rates. Plots (a) of Figure 14 to Figure 19 show that generally there is a non-linear relationship between system-on outdoor air change rates calculated using CONTAM and W_s. Exceptions to these trends are discussed below.

For the Medium Office, the system-on air outdoor change rates calculated using CONTAM are generally constant between ΔT=0 $^{\circ}$C and ΔT=20 $^{\circ}$C and linearly related to ΔT outside this range (Figure 10a). Note that there is also another collection of data points on a line that is symmetrical about ΔT=18 $^{\circ}$C. This separate group of points exists because the system is on from 6 a.m. to 7 a.m., but outdoor air intake is not scheduled until 7 a.m. Thus, from 6 a.m. to 7 a.m., the infiltration rates were affected by the ventilation system airflows and are shifted from the system-off trendline. For the Primary School and Stand-Alone Retail, the system-on outdoor air change rates calculated using CONTAM are generally constant for ΔT<0 and linearly related to ΔT for other values (see Figure 11a and Figure 13a).

The system-on outdoor air change rates calculated using EnergyPlus are fairly constant, except when the economizer is in operation (all buildings except Small Hotel, which had no economizer). This increase in outdoor air change rates is clearly observed in the plots of outdoor air change rate vs. ΔT (plots (b) of Figure 8 to Figure 11 and Figure 13b). No relationship is seen between the system-on outdoor air change rates calculated using EnergyPlus and W_s (plots (b) of Figure 14 to Figure 19). There is also little difference between the system-on outdoor air change rates calculated using the EnergyPlus (CONTAM-equiv.) and EnergyPlus (tight) models *except* for the Small Hotel (Figure 12b) and Stand-Alone Retail (Figure 13b). The dependencies on temperature difference and wind speed, however, were not affected.

Outdoor air change rates vs. weather for system-off hours
Plots (a) of Figure 8 to Figure 13 show that generally there is a linear relationship between system-off outdoor air change rates calculated using CONTAM and ΔT, with the dependence being symmetrical about ΔT=0. This is expected since the stack effect is driven by air density, i.e., air temperature, differences between indoors and outdoors. Plots (a) of Figure 14 to Figure 19 show that generally there is a non-linear relationship between system-off outdoor air change rates calculated using CONTAM and W_s. This dependence is expected since indoor-outdoor pressure differences due to wind are related to the square of the wind speed.

There is approximately a four-fold increase in system-off outdoor air changes rates using the EnergyPlus (CONTAM-equiv.) models compared to the EnergyPlus (tight) models, as reflected in plots (b) of Figure 8 to Figure 13. Since infiltration is a scheduled input in the EnergyPlus models (Section 1.1), the system-off air outdoor change rates calculated using EnergyPlus are constant, whether plotted against ΔT or W_s. Only for the Medium Office and Primary School were the conditions suitable for economizer operation when the HVAC system was scheduled to be off for the EnergyPlus (tight) models. During system-off economizer operation in the Medium Office (Figure 10b) and Primary School (Figure 11b), the outdoor air change rates calculated using EnergyPlus were higher (indicated by black triangles) than the system-off air change rates (indicated by blue x's). Generally, the economizer was in operation for -10 °C <ΔT<20 °C. Note that Figure 10 and Figure 11 show only the outdoor air change rates when the wind speed was less than 2 m/s. In addition to the Medium Office and Primary School, economizers also operated in the Full Service Restaurant for the EnergyPlus (CONTAM-equiv.) models (Figure 8b). During system-off economizer operation, the outdoor air change rates calculated using the EnergyPlus (CONTAM-equiv.) model were higher (indicated by orange triangles) than the system-off outdoor air change rates (indicated by red stars).

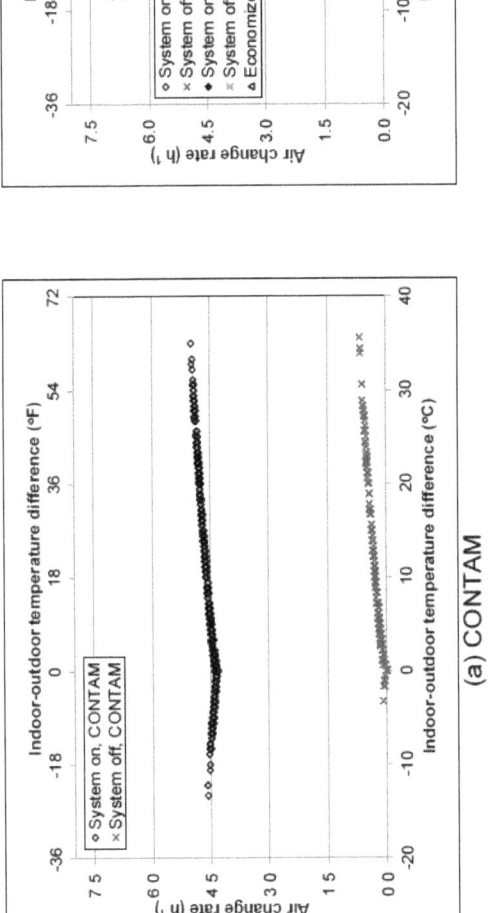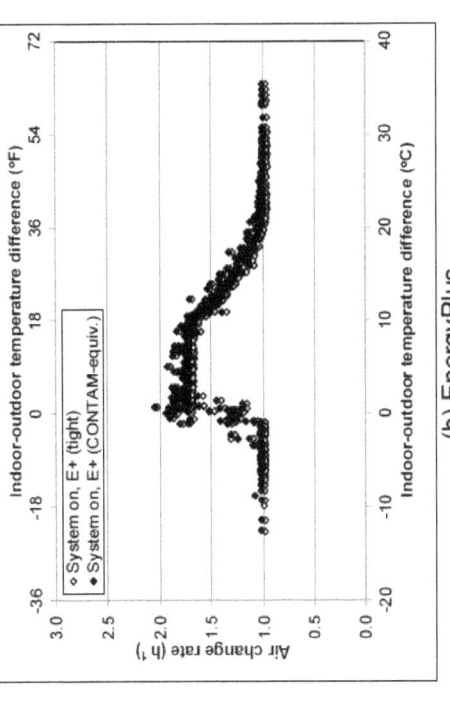

Figure 8 Air change rates as a function of temperature difference (low wind speed) for Full Service Restaurant

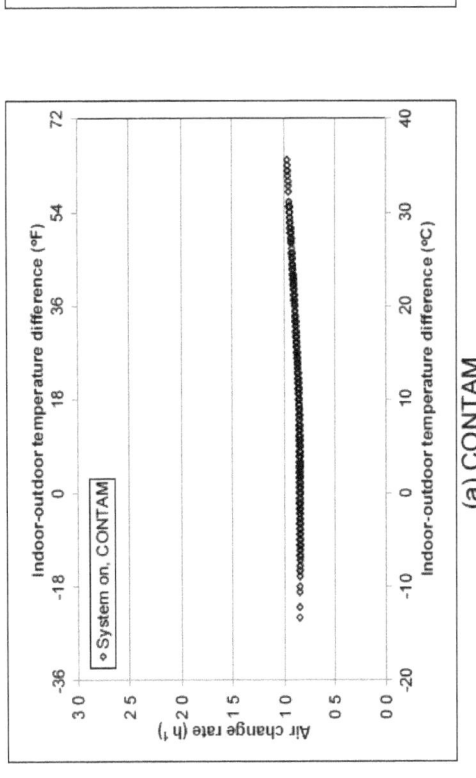

Figure 9 Air change rates as a function of temperature difference (low wind speed) for Hospital

21

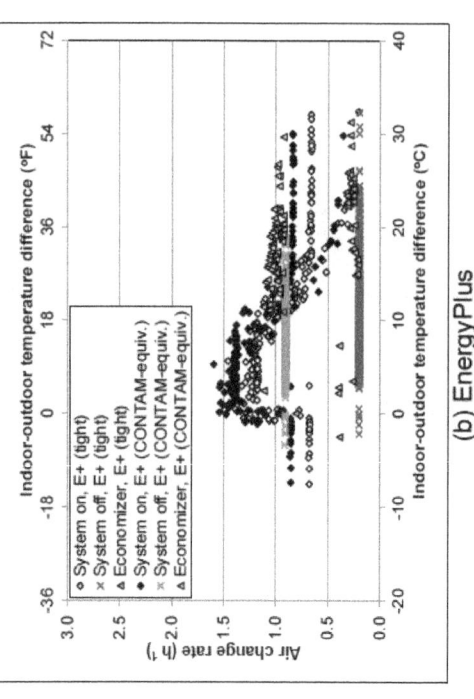

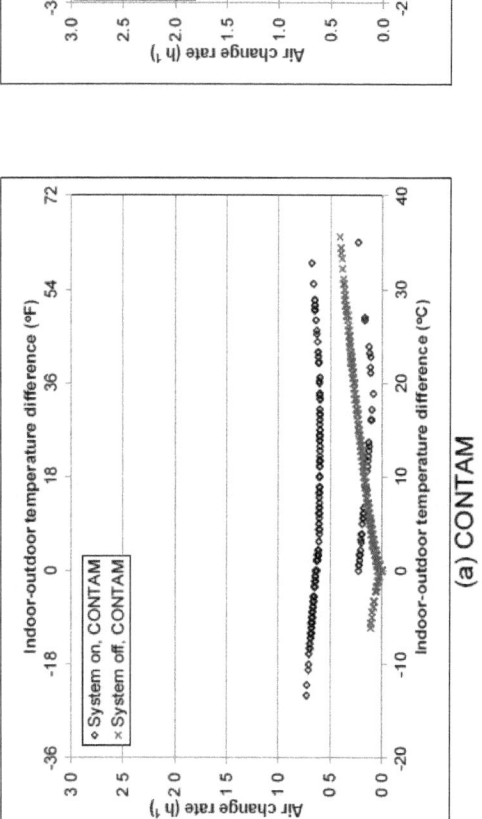

(a) CONTAM

(b) EnergyPlus

Figure 10 Air change rates as a function of temperature difference (low wind speed) for Medium Office

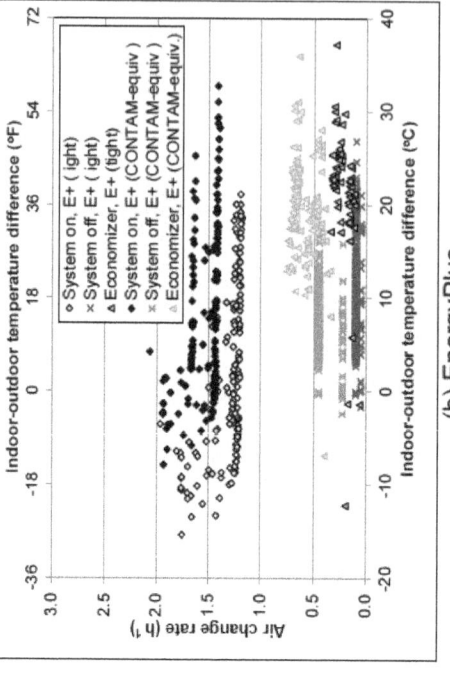

(a) CONTAM

(b) EnergyPlus

Figure 11 Air change rates as a function of temperature difference (low wind speed) for Primary School

22

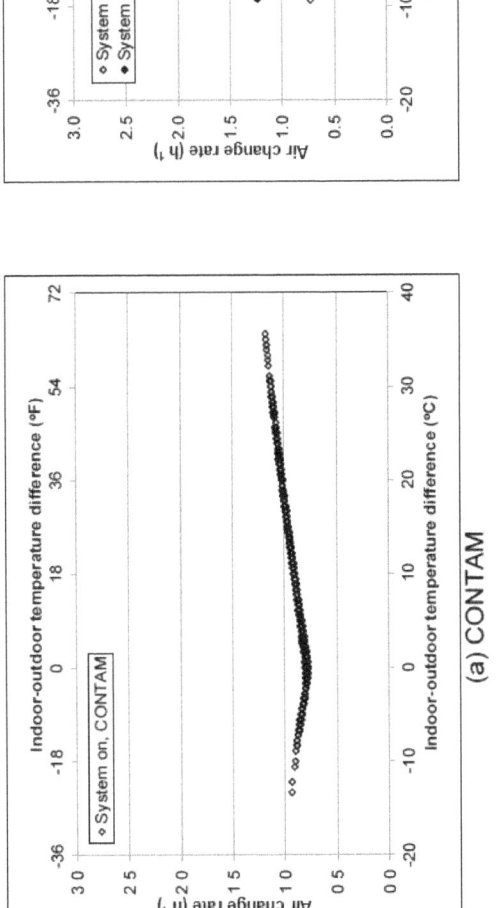

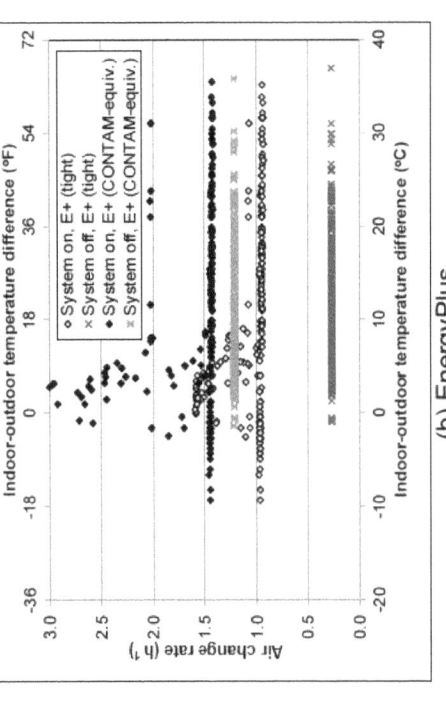

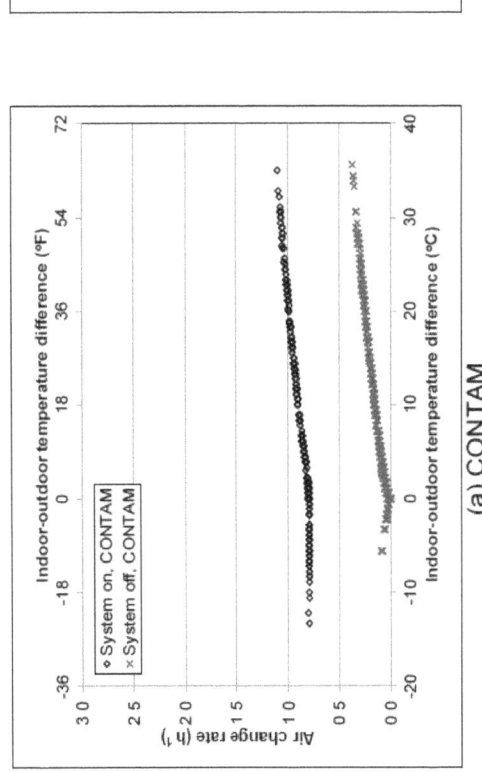

Figure 12 Air change rates as a function of temperature difference (low wind speed) for Small Hotel

Figure 13 Air change rates as a function of temperature difference (low wind speed) for Stand-Alone Retail

23

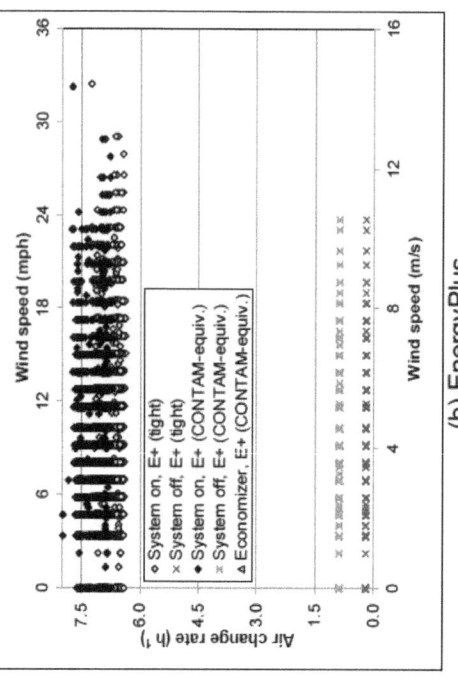

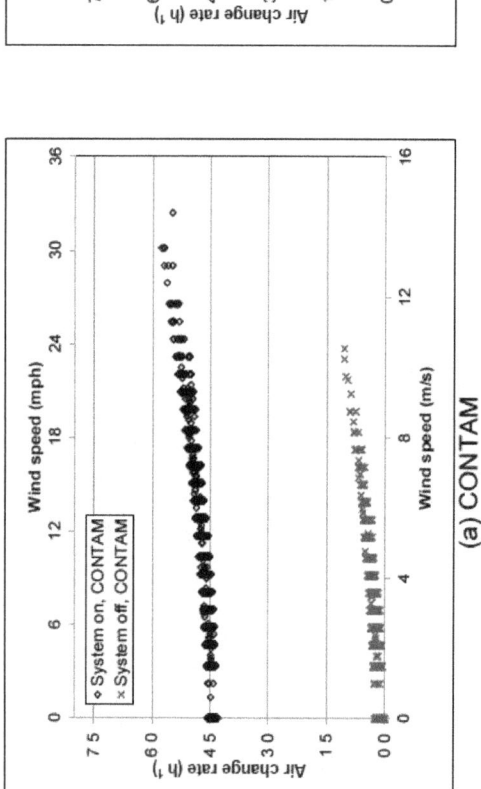

(a) CONTAM

(b) EnergyPlus

Figure 14 Air change rates as a function of wind speed (low ΔT) for Full Service Restaurant

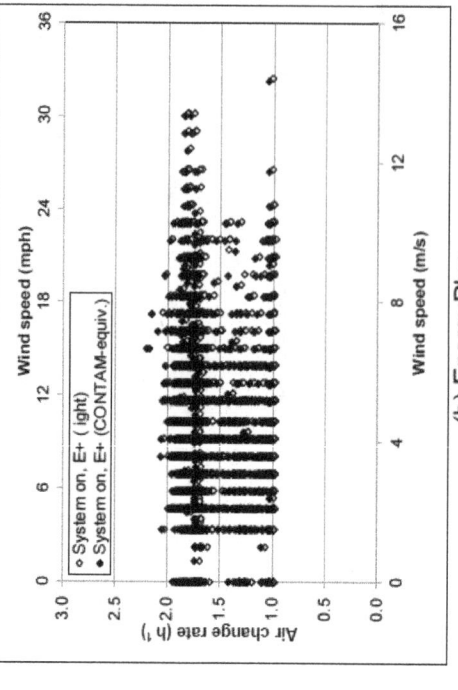

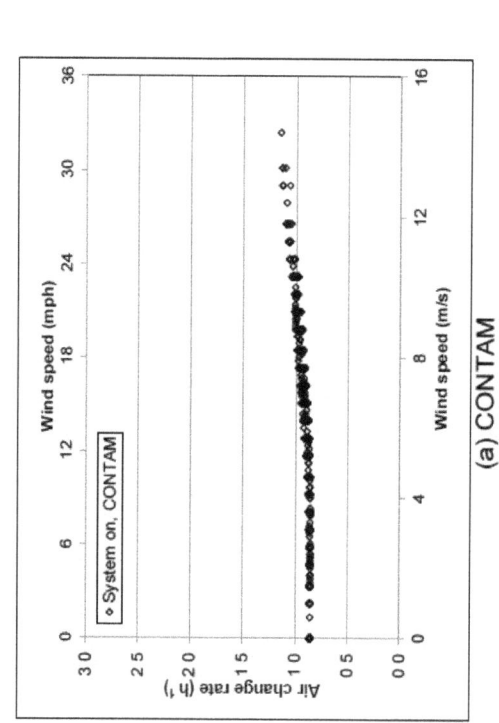

(a) CONTAM

(b) EnergyPlus

Figure 15 Air change rates as a function of wind speed (low ΔT) for Hospital

24

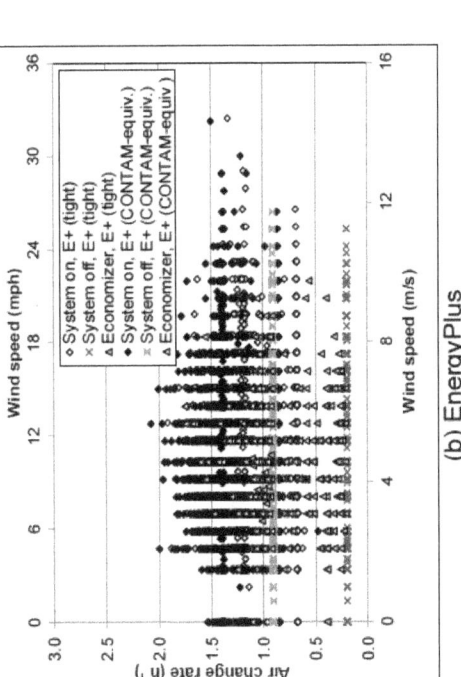

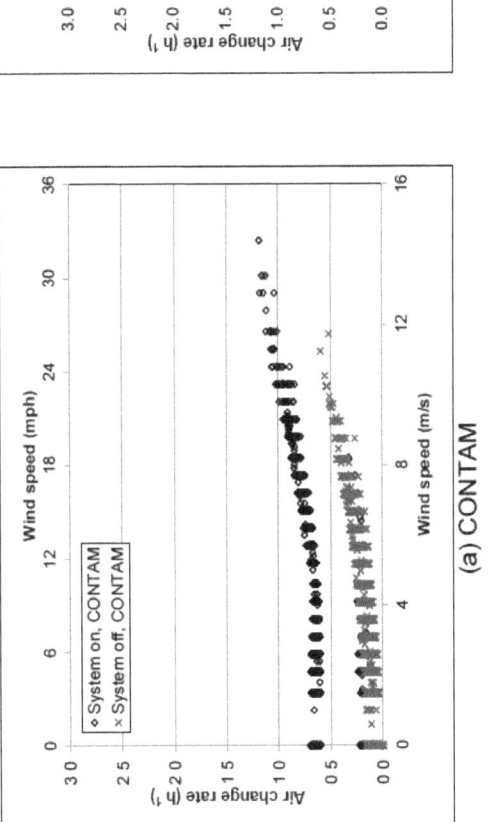

(a) CONTAM

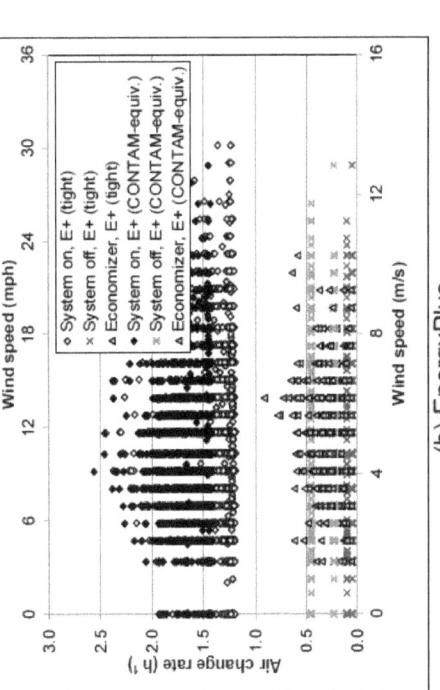

(b) EnergyPlus

Figure 16 Air change rates as a function of wind speed (low ΔT) for Medium Office

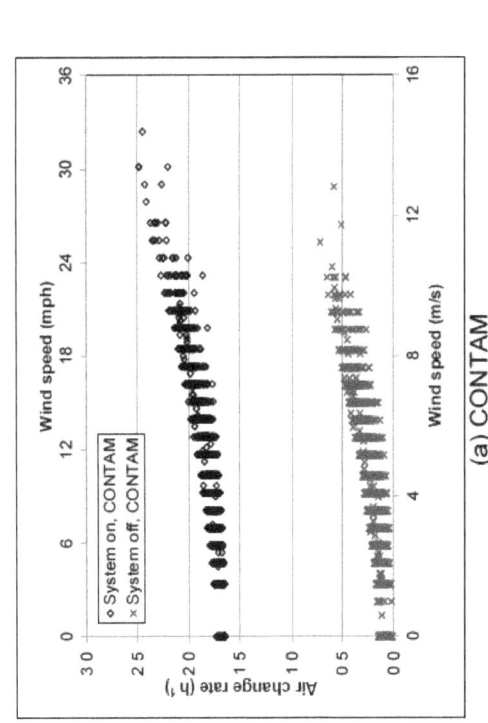

(a) CONTAM

(b) EnergyPlus

Figure 17 Air change rates as a function of wind speed (low ΔT) for Primary School

25

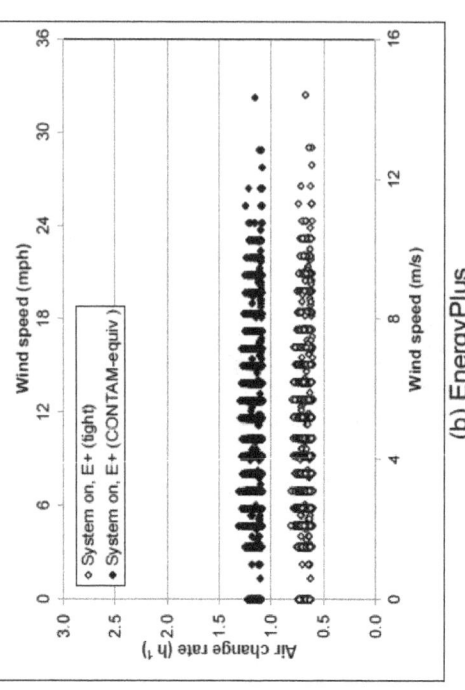

(a) CONTAM

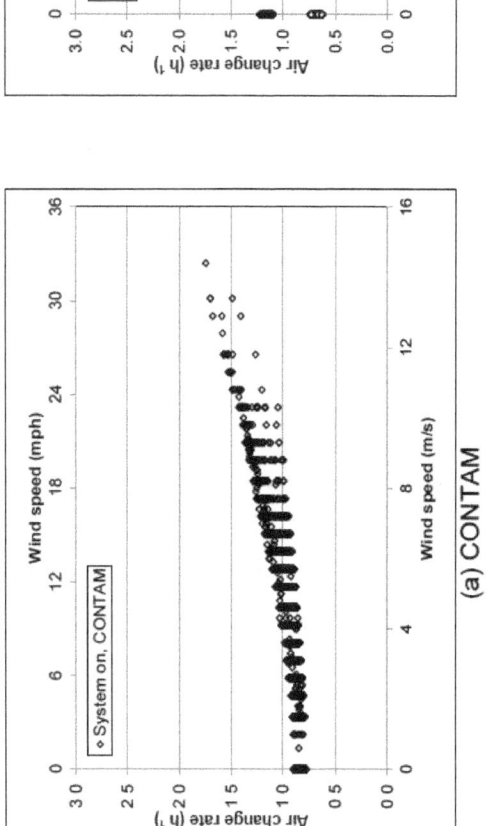

(b) EnergyPlus

Figure 18 Air change rates as a function of wind speed (low ΔT) for Small Hotel

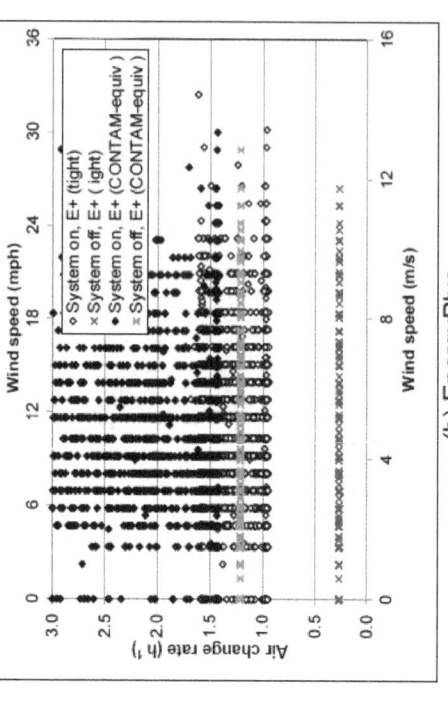

(a) CONTAM

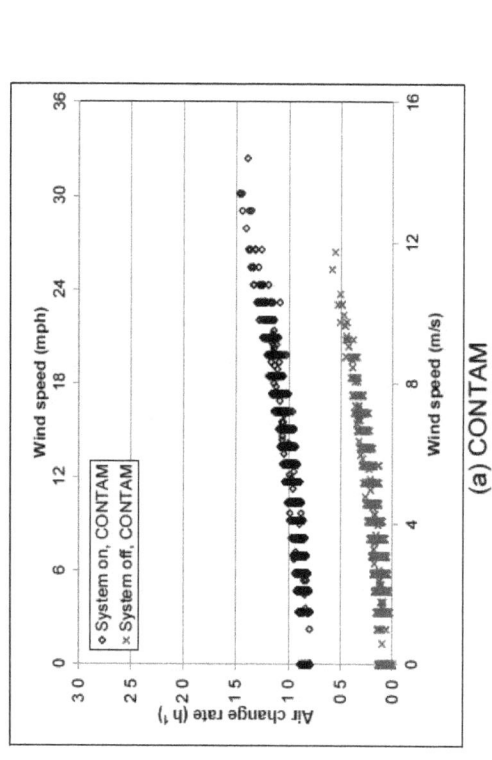

(b) EnergyPlus

Figure 19 Air change rates as a function of wind speed (low ΔT) for Stand-Alone Retail

4.3. Impacts of infiltration on sensible load

As discussed in Section 4.1 and 4.2, there are a significant number of hours out of the year for which the infiltration rates calculated using CONTAM are higher than those calculated using EnergyPlus. This section describes the impact of those higher rates on the sensible heating and cooling loads in the six simulated buildings. The sensible load, q_s in kJ, associated with infiltration is calculated by:

$$q_s = \rho C_P I V (T_{outdoor} - T_{indoor}) \times 1 \text{ h} \qquad (2)$$

where C_P is the specific heat of air in kJ/kg, I is the air change rate due to infiltration in h^{-1}, T_{indoor} is the average indoor temperature of all zones in K, $T_{outdoor}$ is the outdoor temperature in K, and V is the volume of the building in m^3. When $T_{outdoor}$ is less than T_{indoor}, q_s will be negative and thus there is a sensible heating load. Conversely, when $T_{outdoor}$ is greater than T_{indoor}, q_s will be positive and there is a sensible cooling load.

Table 6 lists the total heating and cooling sensible loads due to infiltration in each building in GJ for one year of Chicago weather conditions. These estimates do not account for any other loads, internal or external, or HVAC system effects and efficiencies in meeting these loads. The columns in Table 6 contain the sensible load due to infiltration using the EnergyPlus (tight and CONTAM-equiv. models) infiltration rates, the sensible load using the infiltration rates from CONTAM, and the ratios of the CONTAM infiltration sensible load to the EnergyPlus (tight and CONTAM-equiv. models) infiltration load.

Table 6 Sensible loads due to infiltration

Building	Load	EnergyPlus (tight) (GJ)	EnergyPlus (CONTAM -equiv.) (GJ)	CONTAM (GJ)	Ratio of CONTAM to EnergyPlus (tight)	Ratio of CONTAM to EnergyPlus (CONTAM-equiv.)
Full Service Restaurant	Heating	27	115	128	4.8	1.1
	Cooling	0.4	2	6	13.6	3.0
Hospital	Heating	100	454	898	9.0	2.0
	Cooling	8	36	22	2.7	0.6
Medium Office	Heating	310	1178	611	2.0	0.5
	Cooling	4	16	23	6.6	1.4
Primary School	Heating	221	1083	1248	5.7	1.2
	Cooling	19	18	62	3.2	3.4
Small Hotel	Heating	230	1010	479	2.1	0.5
	Cooling	6	27	22	3.7	0.8
Stand-Alone Retail	Heating	333	1379	559	1.7	0.4
	Cooling	7	32	20	2.6	0.6

In general, the infiltration sensible heating loads are greater than the cooling loads, whether calculated by EnergyPlus or CONTAM. This is to be expected since the simulations were performed using Chicago weather. The infiltration heating loads calculated using CONTAM are about two to nine times higher than the EnergyPlus (tight) results. In the Full Service Restaurant, Hospital, and Primary School, the infiltration sensible heating loads calculated using CONTAM are also higher than the EnergyPlus (CONTAM-equiv.) results. However, rather than two to nine times higher than the EnergyPlus (tight) results, the CONTAM results were only 10 % to two times higher than the EnergyPlus (CONTAM-equiv.) results. For the Medium Office, Small Hotel, and Stand-Alone Retail, the infiltration load calculated using CONTAM is 50 % to 60 % lower than the EnergyPlus (CONTAM-equiv.) results. As discussed in Section 4.1, in these three buildings, the CONTAM system-on or system-off infiltration rates, in some cases both, were lower than the EnergyPlus (CONTAM-equiv.) results. Note that these loads only accounted for the heating and cooling of the infiltration air and do not include latent loads, system effects or internal loads.

The infiltration cooling loads calculated using CONTAM were about three to 10 times higher than the EnergyPlus (tight) results. However, for the Full Service Restaurant, the cooling load was 14 times higher than the EnergyPlus (tight) results, largely due to the fact that the EnergyPlus (tight) infiltration sensible cooling load was very small, 0.4 GJ. In the Full Service Restaurant, Medium Office, and Primary School, the infiltration sensible cooling loads calculated using CONTAM were also higher than the EnergyPlus (CONTAM-equiv.) results. However, rather than three to 10 times higher than the EnergyPlus (tight) results, the CONTAM results were only 40 % to three times higher than the EnergyPlus (CONTAM-equiv.) results. For the Hospital, Small Hotel, and Stand-Alone Retail, the infiltration sensible cooling load calculated using CONTAM was 20 % to 40 % lower than the EnergyPlus (CONTAM-equiv.) results. As discussed in Section 4.1, both the CONTAM system-on and system-off infiltration rates for the Small Hotel and Stand-Alone Retail were lower than the EnergyPlus (CONTAM-equiv.) results.

For all of the buildings, except the Hospital, the differences in calculated infiltration sensible load between the CONTAM and EnergyPlus (tight) models were 10 % to 60 % of the total energy consumption. For the Hospital, the differences in calculated infiltration sensible load were only 1 % to 5 % of the total energy consumption because the energy consumption for the Hospital is the highest among the six simulated buildings. Although the infiltration sensible loads calculated using the EnergyPlus (CONTAM-equiv.) models were closer to the CONTAM values, there were still 10 % to 50 % differences between CONTAM and EnergyPlus (CONTAM-equiv.) results.

Note that the values in Table 6 are total sensible loads over one year and that the differences between the CONTAM- and EnergyPlus loads are more significant for individual hours when the weather conditions lead to higher infiltration rates. All of these differences point out the importance of accurate modeling of building airflow in building energy simulation.

5. CONTAMINANT SIMULATION RESULTS

This section presents the contaminant simulation results for system-on hours only, during which the majority of the occupants are present. The distribution of daily average concentrations of all contaminants, distribution of daily peak CO_2 concentrations, and distribution of daily peak volatile organic compounds (VOC) concentrations are plotted in Figure 20 to Figure 37 for the selected zones listed in Table 7 for each of the six buildings. Detailed results can be found in Appendix B. Table 7 also lists the total maximum occupancy in those zones, the total zone occupancy divided by the floor area of the zones, and the average outdoor air intake per person for the zones. For the Full Service Restaurant and Stand-Alone Retail, the only zone for which results are not presented is for the restroom because it is not generally occupied. For the remaining buildings, the zones were selected to represent various occupancy types and densities within the building. Zones were also selected based on location within the building to represent different exposures to weather conditions. For instance, in the Medium Office, both the West and South Perimeter zones were selected since they would be exposed to different wind conditions. In the Medium Office and Small Hotel, similar zones were selected on each floor in order to observe the differences in contaminant concentrations due to elevation. As seen in Table 7, the highest occupancy density in the Full Service Restaurant, while the Hospital and Medium Office are almost a factor of ten lower. The outdoor air intake rates per person are similar for all the buildings, except for the Hospital, for which they are much higher due to the higher ventilation requirements in healthcare facilities. Table 8 lists the minimum and maximum daily average and daily peak indoor concentration for each contaminant based on the zones listed in Table 7. Thus, the values in Table 8 do not reflect overall minimum and maximum for the entire building but only for those zones. Detailed minimums, maximums, means and standard deviations of the indoor concentrations for each zone in each building are in Appendix B. Figure 20 to Figure 25 show the distribution frequency of CO_2 concentration for each selected zone in each of the six buildings for the entire year. Figure 26 to Figure 31 show the distribution frequency of VOC concentration for each selected zone. Figure 32 to Figure 37 show the distribution frequency of ozone and PM 2.5 concentration for each selected zone.

The Full Service Restaurant and Primary School are among the buildings with the highest indoor ozone and PM 2.5 concentrations as shown in Table 8, but the differences are not exceptionally large. The only source of ozone and PM 2.5 is the outdoor air, and their indoor deposition rates are related to zone size. Thus, the slightly higher indoor ozone and PM 2.5 concentrations in these two buildings occur primarily due to the higher air change rates that bring in more ozone and PM 2.5. In contrast, the Hospital and Medium Office are among the buildings with the lowest indoor ozone and PM 2.5 concentrations. This is due primarily to the lower air change rates that bring in less ozone and PM 2.5.

Table 8 also shows that the Hospital and Medium Office have the highest indoor VOC concentrations. Because the VOC sources are area-based, the primary factor in determining the indoor concentrations are the outdoor air intake rates per unit floor area. The Hopstial and Medicum Office have the lowest values of the outdoor air intake per floor area, leading to the highest VOC concentrations. In contrast, the Primary School, Small Hotel, and Stand-Alone Retail have higher outdoor air intake rates per floor area, leading to lower indoor VOC concentrations.

Table 7 Selected zones for which contaminant concentration results reported

Building	Selected zones		Total max. occupancy in selected zones[1]	Max. occupancy per floor area (per 100 m²)[1]	Average outdoor air intake (L/s•person)[2]
Full Service Restaurant	Dining	Kitchen	274	55	10.0
Hospital	1F ER Exam 3 1F ER Nurse's Station 1F Lobby 2F ICU 2F ICU Patient Rm 3 2F Operating Rm 2	3F Lab 3F Nurse's Station Lobby 3F Patient Rm 3 3F Patient Rm 4 5F Dining 5F Office 2	259	4	27.1
Medium Office	1-3F Core Zone 1-3F West Perimeter	1-3F South Perimeter	213	6	11.9
Primary School	Cafeteria Gym Library/Media Classroom	Offices Pod 1 Corner Classroom 1 Pod 1 Multiple Classroom 1	591	29	9.4
Small Hotel	Front Lounge Meeting Room Guest 209-212 Guest 309-312	Guest 409-412 Guest 215-218 Guest 315-318 Guest 415-318	129	13	9.2
Stand-Alone Retail	Back Space Core Retail	Front Retail Point of Sale	321	14	9.8

Notes:
1. Maximum occupancy and occupancy per floor area are based only on the selected zones.
2. Values are sum of L/s supplied to all of the selected zones divided by the sum of the number of occupants in the selected zones.

Table 8 Summary of calculated contaminant concentrations

Full Service Restaurant	Daily average contaminant concentrations		Daily peak contaminant concentrations	
	Min.	Max.	Min.	Max.
CO_2, mg/m^3	1020	1579	1352	2433
Ozone, μg/m^3	3	62	5	86
PM 2.5, μg/m^3	1	42	2	61
VOC, μg/m^3	27	62	31	343

Hospital	Daily average contaminant concentrations		Daily peak contaminant concentrations	
	Min.	Max.	Min.	Max.
CO_2, mg/m^3	749	991	759	1145
Ozone, μg/m^3	1	40	1	58
PM 2.5, μg/m^3	<1	24	1	33
VOC, μg/m^3	38	242	42	243

Medium Office	Daily average contaminant concentrations		Daily peak contaminant concentrations	
	Min.	Max.	Min.	Max.
CO_2, mg/m^3	826	1219	887	1416
Ozone, μg/m^3	1	34	2	57
PM 2.5, μg/m^3	<1	23	1	32
VOC, μg/m^3	75	291	104	812

Primary School	Daily average contaminant concentrations		Daily peak contaminant concentrations	
	Min.	Max.	Min.	Max.
CO_2, mg/m^3	793	1279	820	1683
Ozone, μg/m^3	2	74	5	98
PM 2.5, μg/m^3	<1	52	1	68
VOC, μg/m^3	15	172	34	1186

Small Hotel	Daily average contaminant concentrations		Daily peak contaminant concentrations	
	Min.	Max.	Min.	Max.
CO_2, mg/m^3	759	1054	847	1363
Ozone, μg/m^3	1	61	1	89
PM 2.5, μg/m^3	<1	41	1	60
VOC, μg/m^3	16	282	22	306

Stand-Alone Retail	Daily average contaminant concentrations		Daily peak contaminant concentrations	
	Min.	Max.	Min.	Max.
CO_2, mg/m^3	741	1173	757	1486
Ozone, μg/m^3	1	43	2	68
PM 2.5, μg/m^3	<1	32	1	45
VOC, μg/m^3	20	180	24	436

Note: The "Min." and "Max." values only apply to the zones in each building for which contaminant concentrations are reported (see Table 7 for selected zones).

Figure 20 to Figure 25 show the average and peak CO_2 concentrations for the six simulated buildings. The differences in these values among the selected zones reflect variations in occupant density and outdoor air intake per person. For example, Figure 24a shows that the guestrooms in the Small Hotel have higher daily average CO_2 concentrations than the Front Lounge and Meeting Room even though the maximum occupancies in the guestrooms are less. This is because the Front Lounge and Meeting Room are occupied less during the day than the guestrooms, resulting in less total CO_2 being generated in those spaces. Further, the Front Lounge and Meeting Room also have higher outdoor air requirements. Figure 24b shows that the daily peak CO_2 concentrations in the Front Lounge and Meeting Room are larger than those in the guestrooms because they have the largest maximum occupancy of the selected zones.

In general, Figure 26 to Figure 31 show that the distribution of indoor VOC concentrations among selected zones is similar in most buildings. Larger differences are seen for the Hospital (Figure 27) and Small Hotel (Figure 30). In the Hospital, the first floor ER Exam 3 and second floor OR 2 have lower daily average and peak VOC concentrations compared to the other selected zones. This is due to less VOC being generated inside these smaller zones, and the outside air requirements in these zones are among the highest of the selected zones. Figure 30 shows that the daily average and peak VOC concentrations in the Front Lounge and Meeting Room in the Small Hotel are less than those in the guestrooms. This is the case because the outside air requirements in these zones are much higher than in the guestrooms, even though the Front Lounge and Meeting Room are larger than the guestrooms in terms of floor area.

In general, Figure 32 to Figure 37 show that the distribution of indoor ozone and PM 2.5 concentrations among selected zones is similar within each simulated building. This was to be expected since (1) the only source of ozone and PM 2.5 was the outdoors, and (2) differences in how much ozone and PM 2.5 enter these zones from the outdoors (air change rates based on zone volume) were counterbalanced by differences in deposition of these contaminants also based on zone volume. Indoor ozone and PM 2.5 concentrations are lower than the outdoor concentrations for all simulated buildings due to deposition and filtration.

Based on the 2010 ozone data from the EPA AQS database (EPA 2011a) used in these simulations for Chicago, there were only 4 hours during the year (8760 total hours) for which the outdoor ozone level exceeded the National Ambient Air Quality Standards (NAAQS) limit of $150 \mu g/m^3$ (EPA 2011b). The World Health Organization (WHO) indoor guideline for ozone is $100 \mu g/m^3$ (WHO 2005). There were no hours for which the indoor ozone level exceeded $100 \mu g/m^3$ in any of the selected zones in the buildings. Table 8 shows that of the selected zones in the simulated buildings, the maximum daily peak ozone concentration was $98 \mu g/m^3$.

Based on the 2010 PM 2.5 data from the EPA Air Quality Standard (AQS) database (EPA 2011a) used in these simulations for Chicago, there were 868 hours during the year for which the outdoor PM 2.5 level exceeded the NAAQS limit of $35 \mu g/m^3$ (EPA 2011b). The WHO indoor guidelines for PM 2.5 are $10 \mu g/m^3$, $15 \mu g/m^3$, $25 \mu g/m^3$, or $35 \mu g/m^3$ based on the level of risk tolerance (WHO 2005). The number of hours the indoor PM2.5 level exceeded the WHO limit of $35 \mu g/m^3$ was different in each simulated building. In the Full Service Restaurant, there were 122 hours for which the indoor PM 2.5 level exceeded $35 \mu g/m^3$ in the Kitchen. In the Primary School, there were 220 hours for which the indoor PM 2.5 level exceeded $35 \mu g/m^3$ in the

Cafeteria. The maximum number of hours for which the indoor PM 2.5 level exceeded the WHO limit in the Small Hotel was 142 (in the Meeting Room). The maximum number of hours for which the indoor PM 2.5 level exceeded the WHO limit in the Stand-Alone Retail was 16 (in the Back Space). There were no hours for which the indoor PM 2.5 level exceeded the WHO limit in the Medium Office and Hospital.

In the Medium Office and Small Hotel, similar zones were selected on multiple floors in order to observe the differences in contaminants due to elevation. For the indoor contaminants studied in this report, the distributions of concentration remained relatively the same independent of zone location or elevation. This result was perhaps due to the fact that the Medium Office was only three stories and the Small Hotel was only four stories. In taller buildings with a stronger stack effect, there might be more significant concentration gradients with elevation.

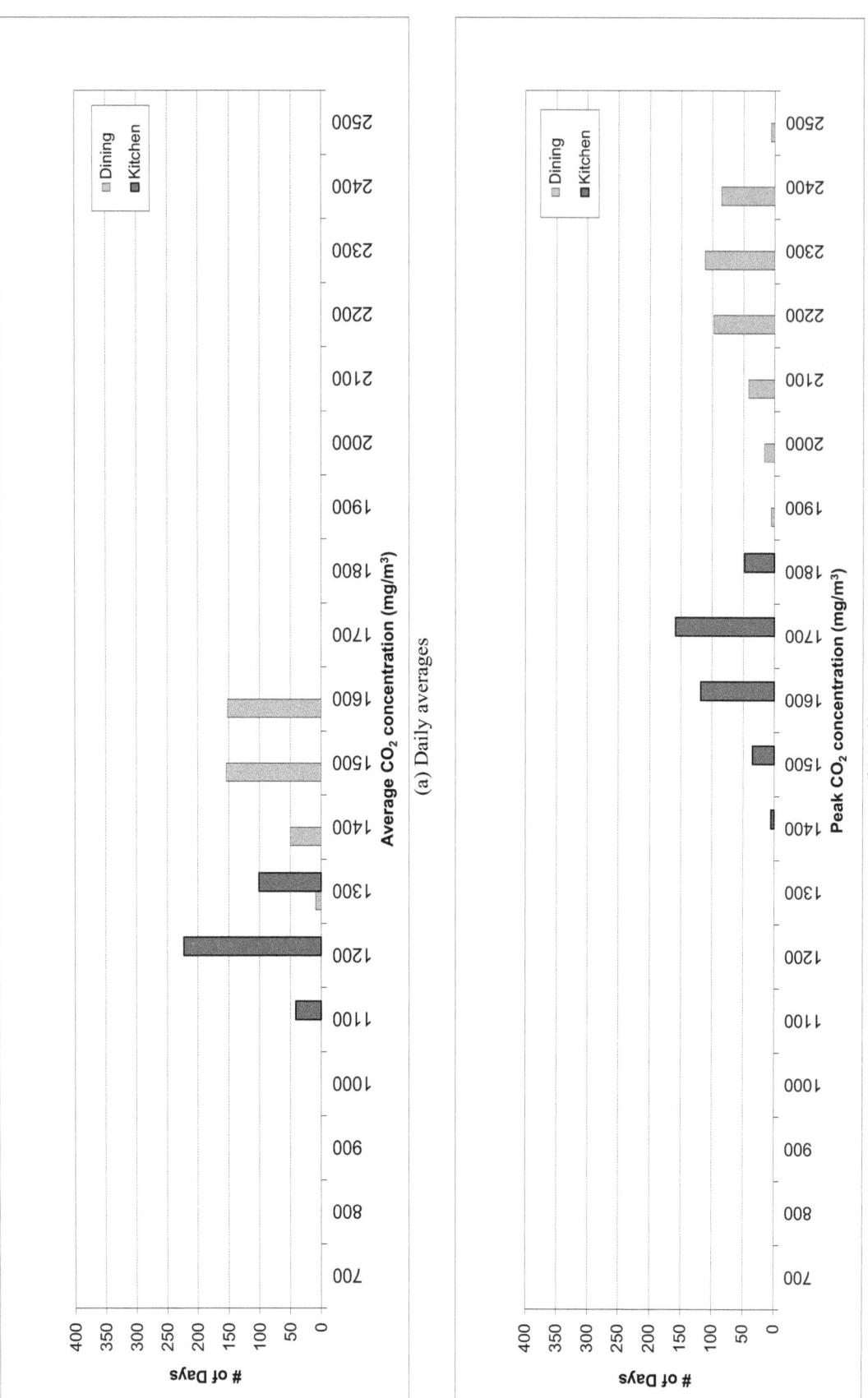

(a) Daily averages

(b) Daily peaks

Figure 20 Frequency distribution of simulated CO$_2$ concentration for Full Service Restaurant

34

(a) Daily averages

(b) Daily peaks

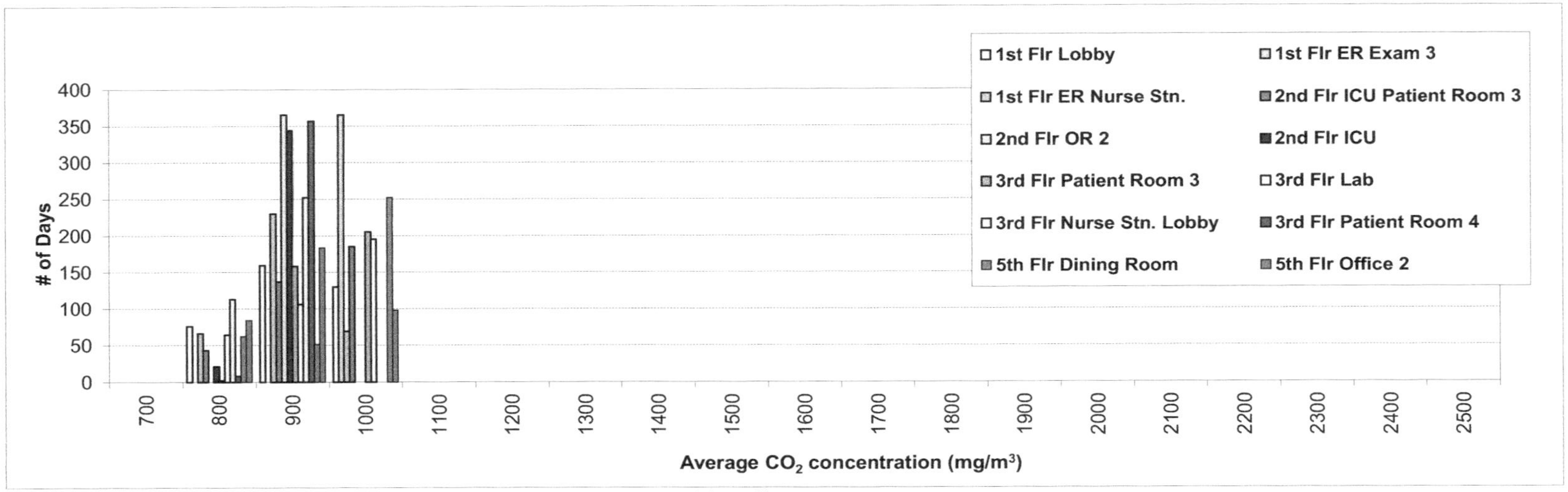

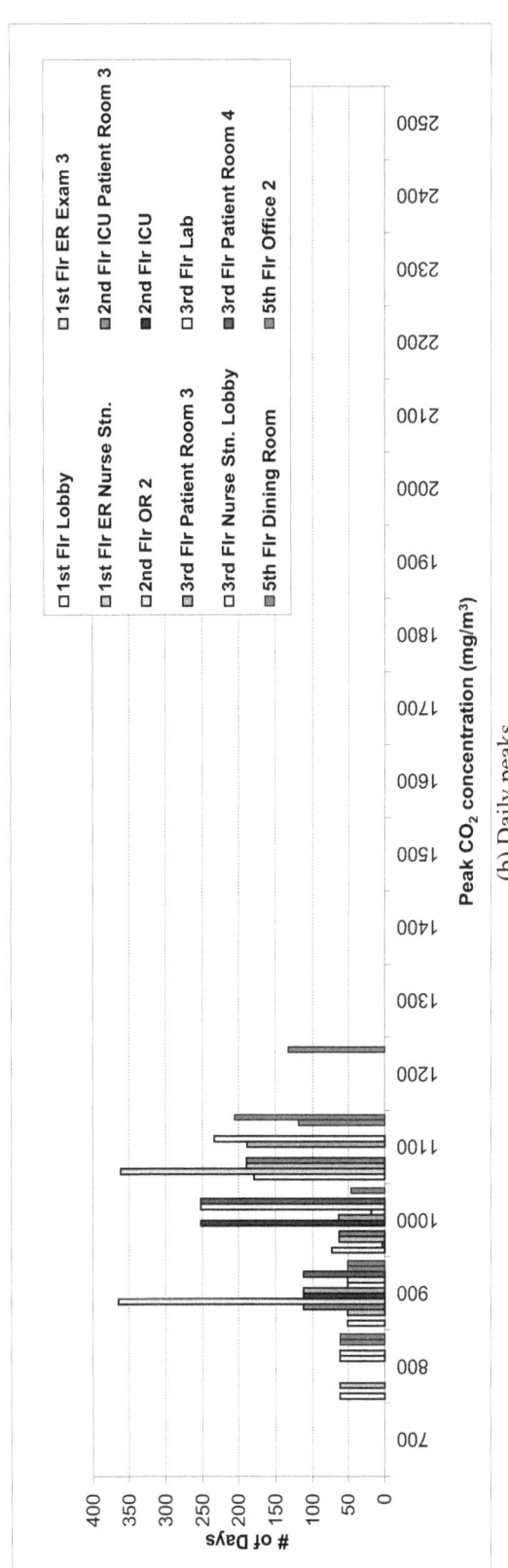

Figure 21 Frequency distribution of simulated CO_2 concentration for Hospital

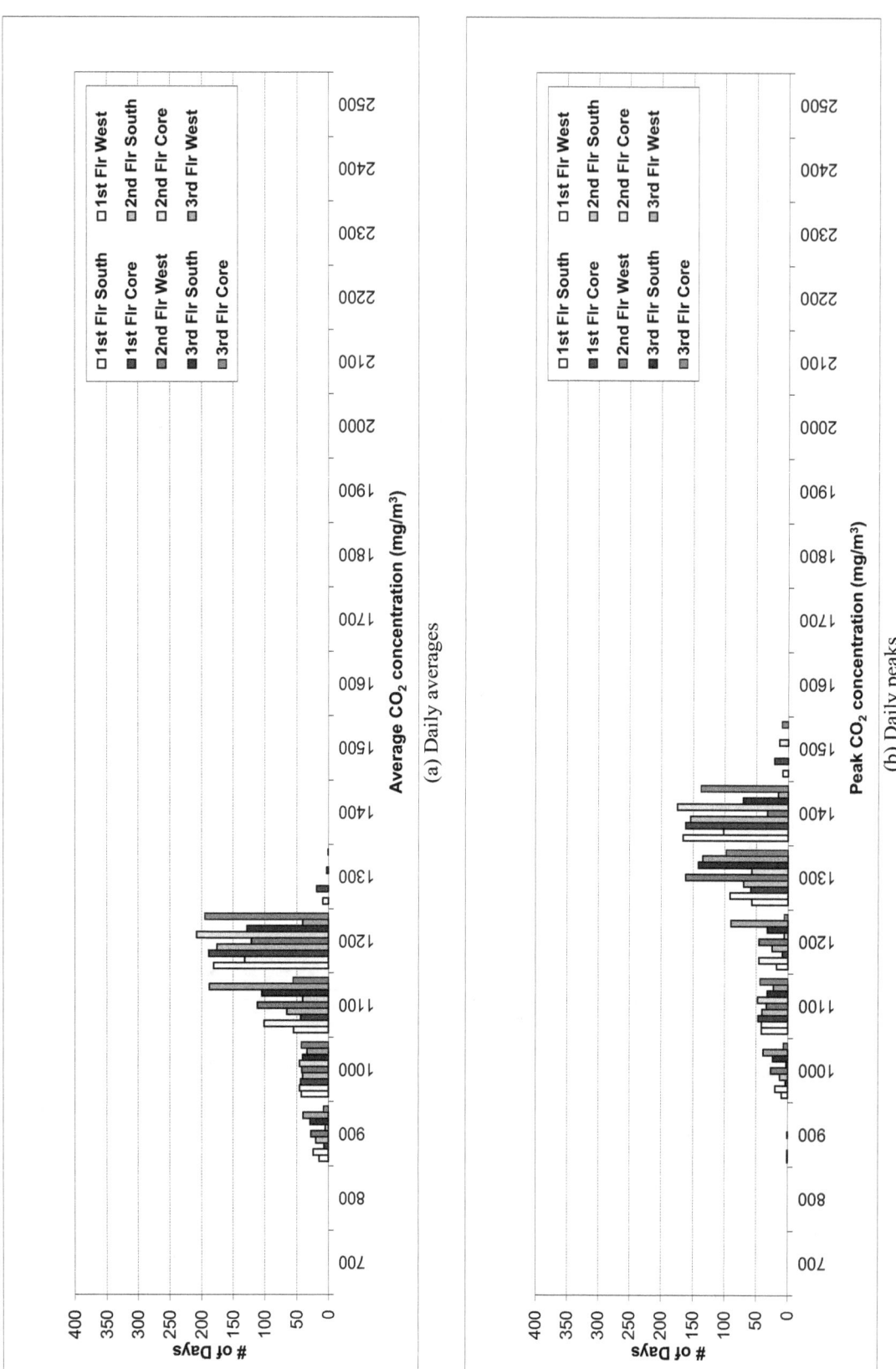

(a) Daily averages

(b) Daily peaks

Figure 22 Frequency distribution of simulated CO$_2$ concentration for Medium Office

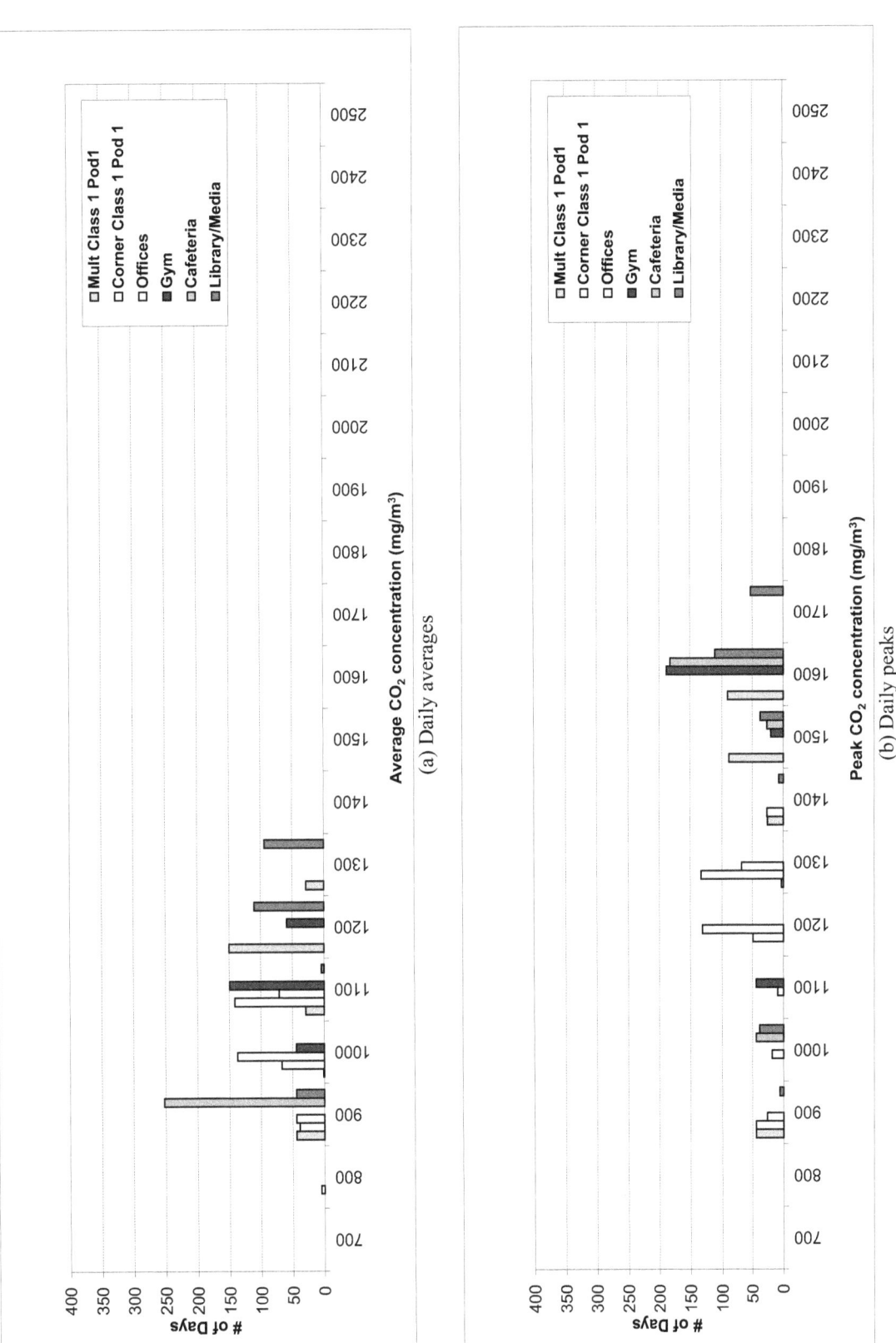

(a) Daily averages

(b) Daily peaks

Figure 23 Frequency distribution of simulated CO₂ concentration for Primary School

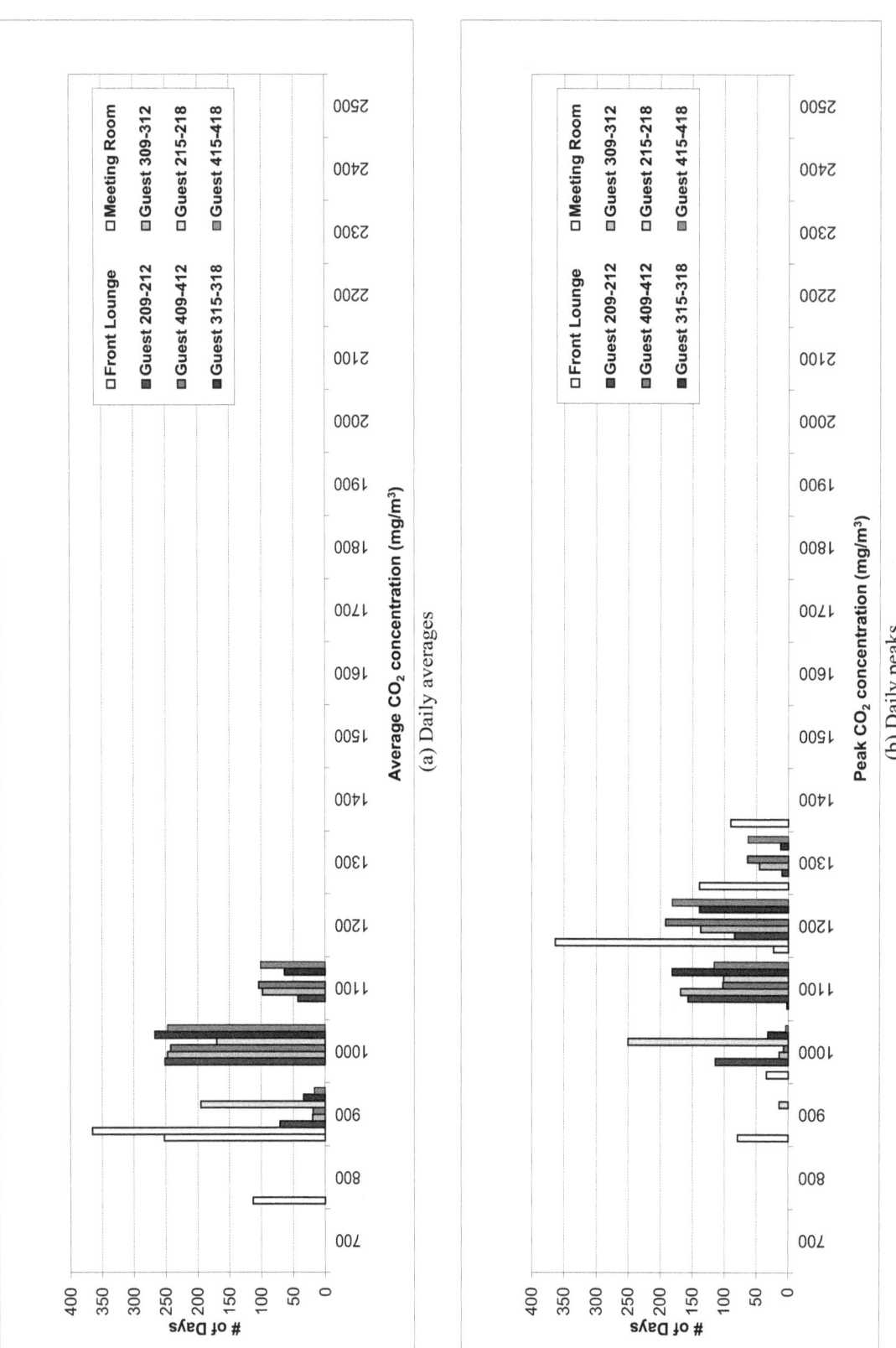

(a) Daily averages

(b) Daily peaks

Figure 24 Frequency distribution of simulated CO$_2$ concentration for Small Hotel

38

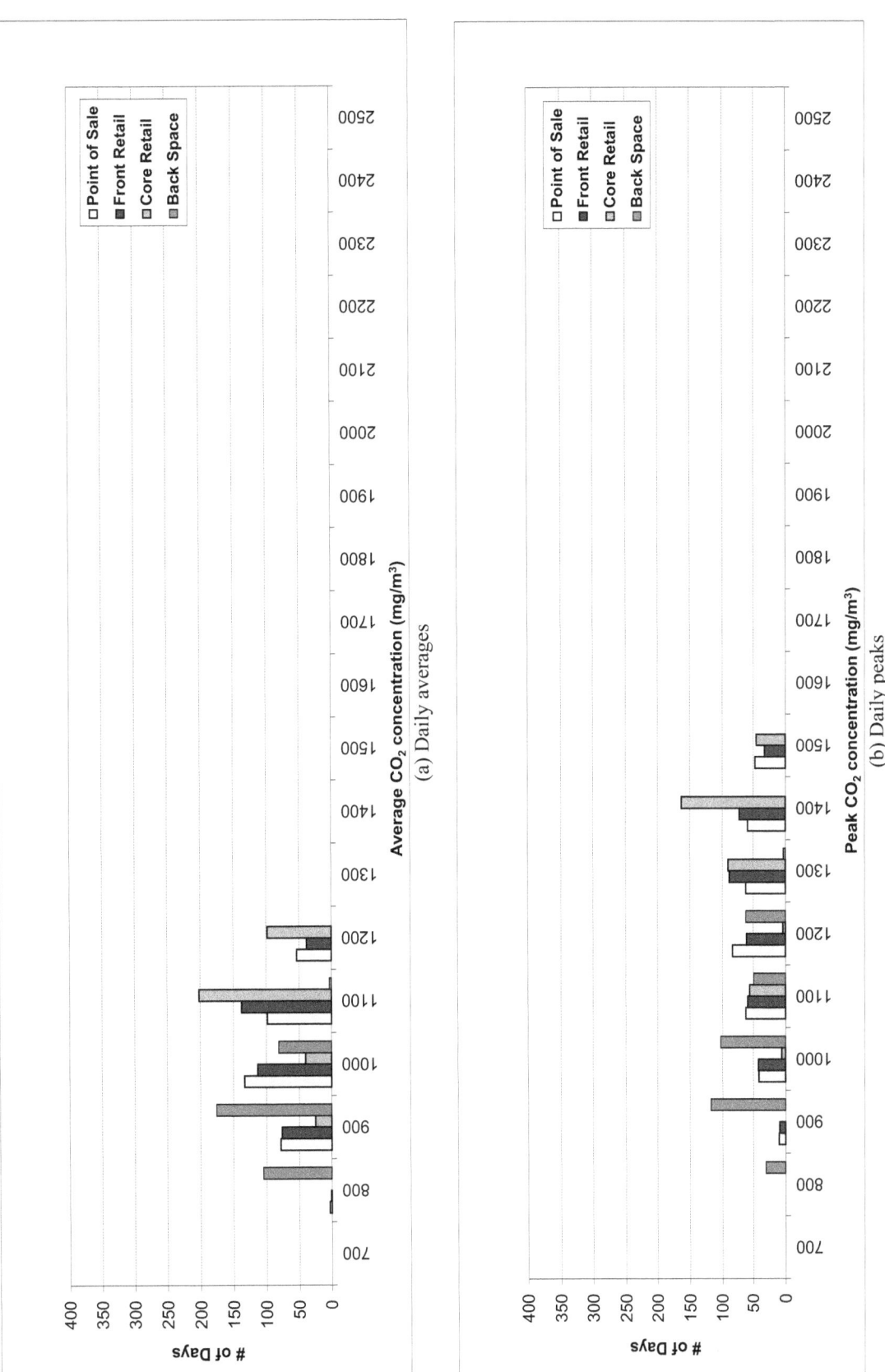

Figure 25 Frequency distribution of simulated CO_2 concentration for Stand-Alone Retail

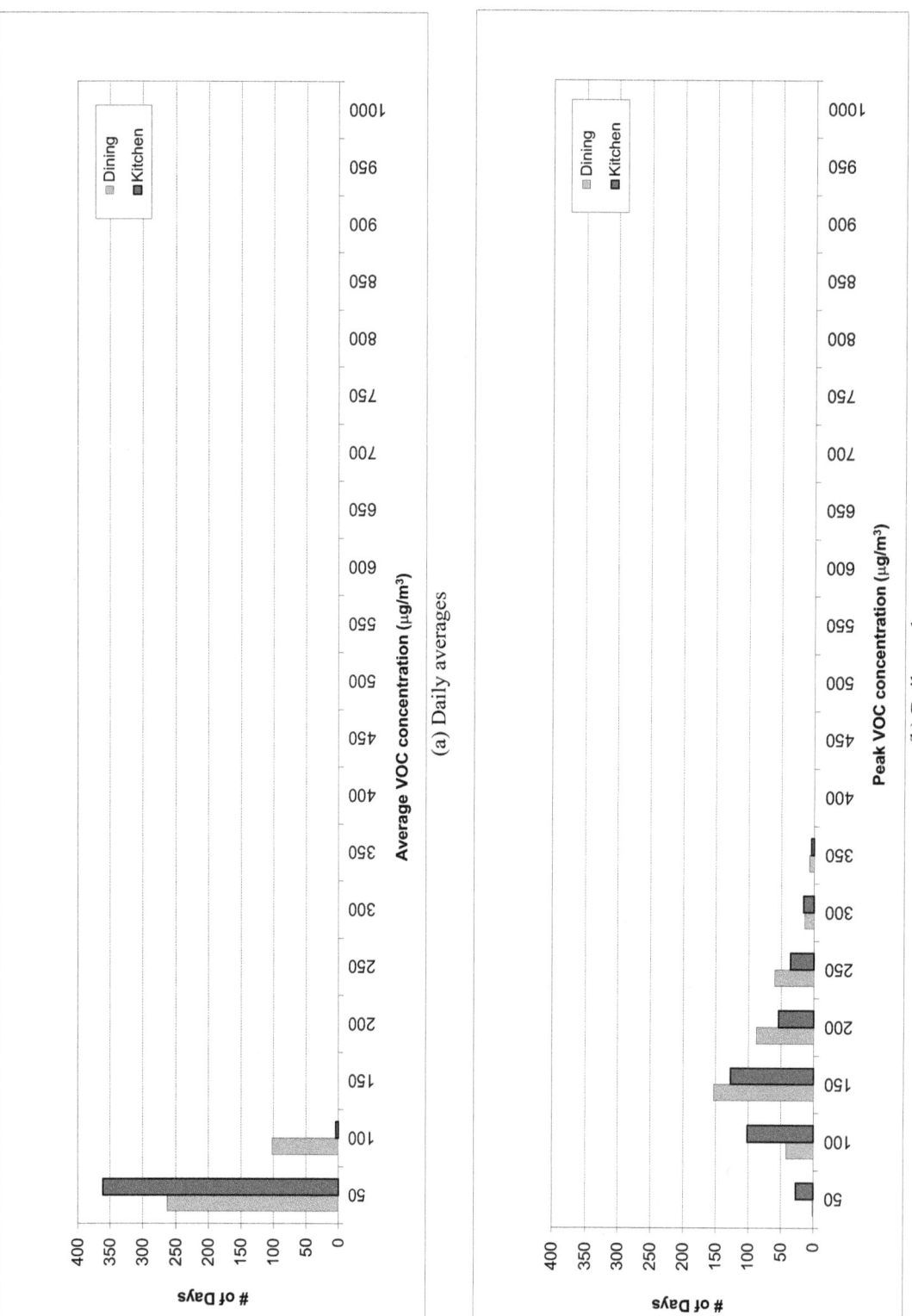

(a) Daily averages

(b) Daily peaks

Figure 26 Frequency distribution of simulated VOC concentration for Full Service Restaurant

40

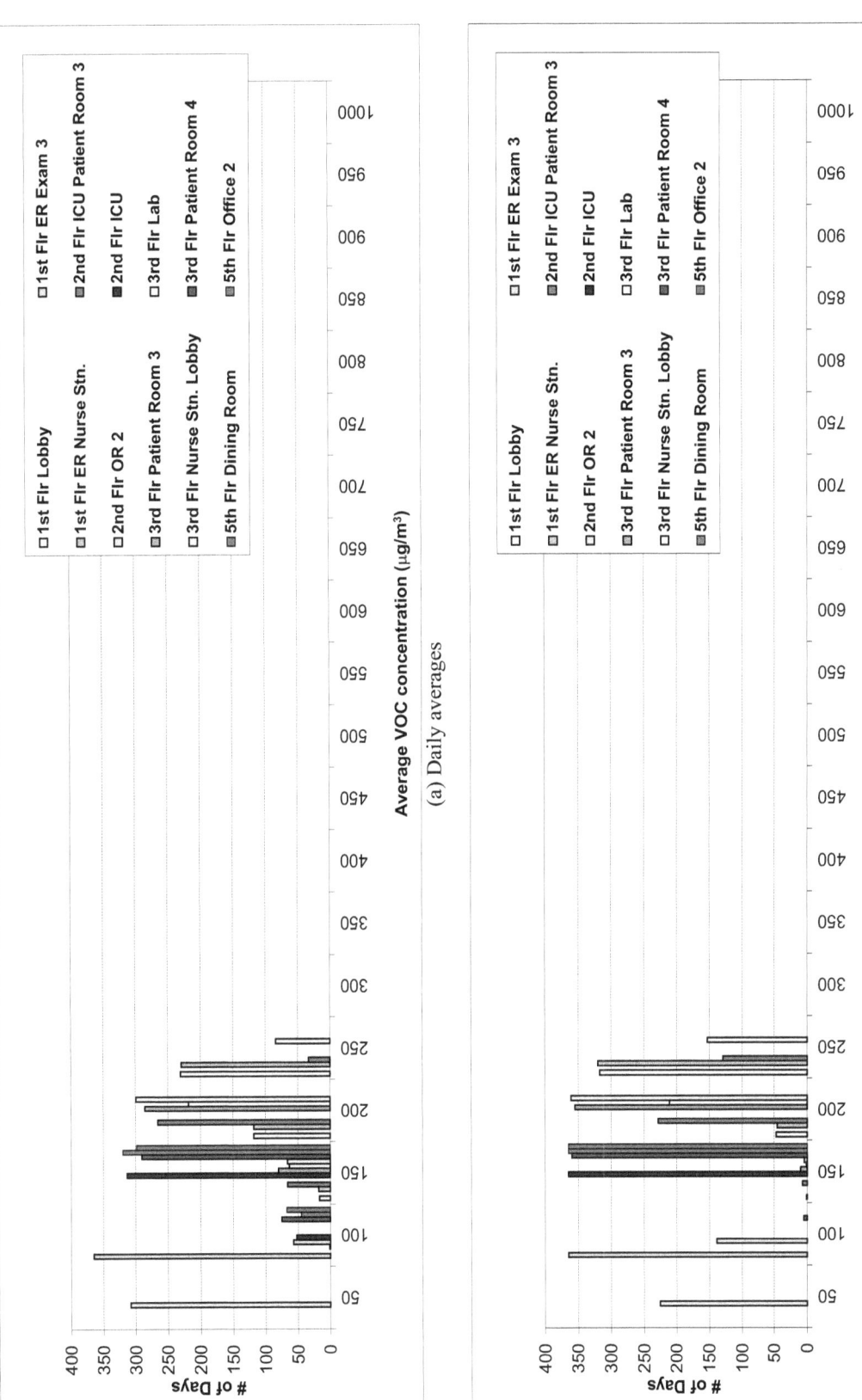

(a) Daily averages

(b) Daily peaks

Figure 27 Frequency distribution of simulated VOC concentration for Hospital

41

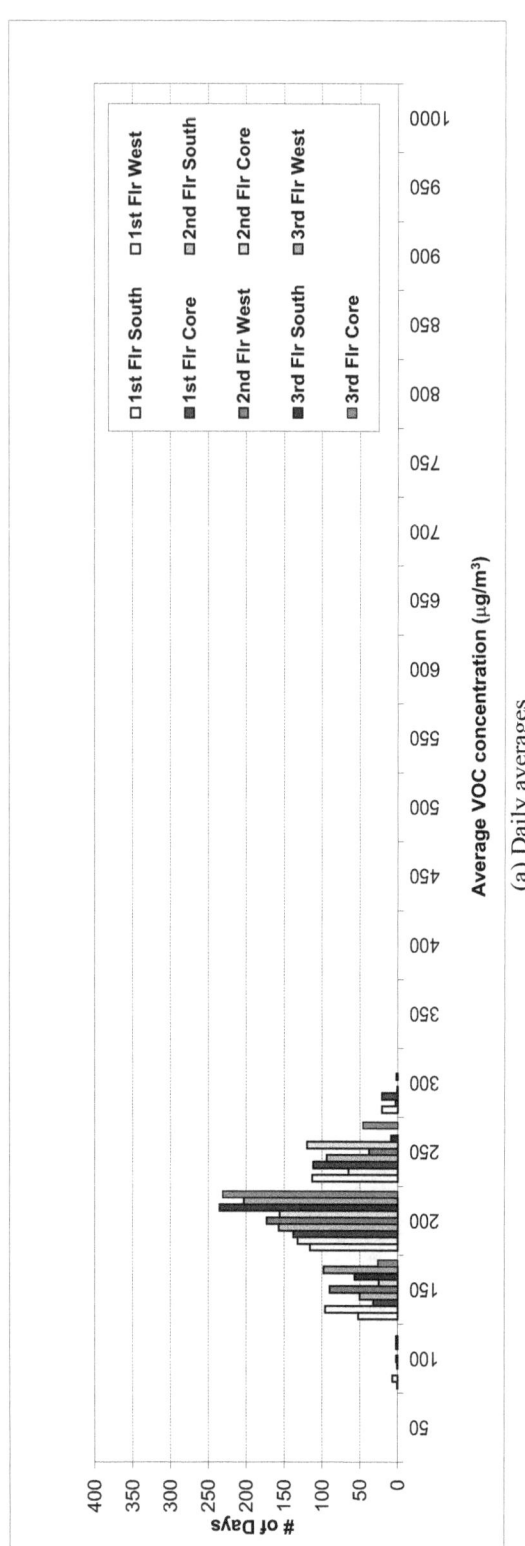

(a) Daily averages

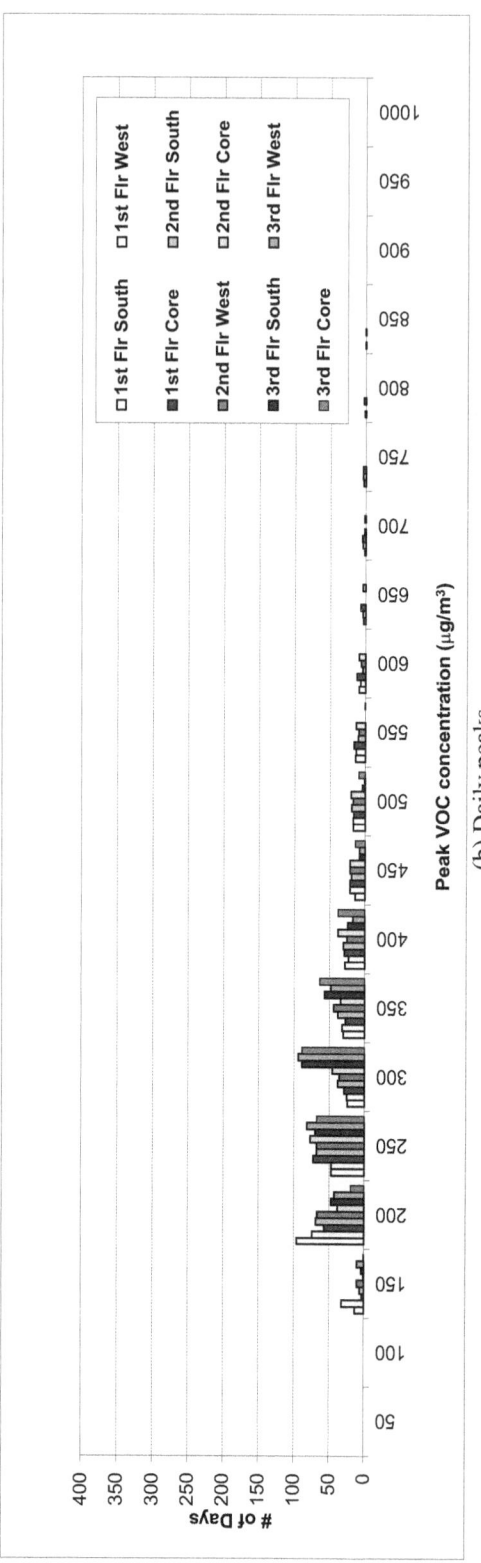

(b) Daily peaks

Figure 28 Frequency distribution of simulated VOC concentration for Medium Office

42

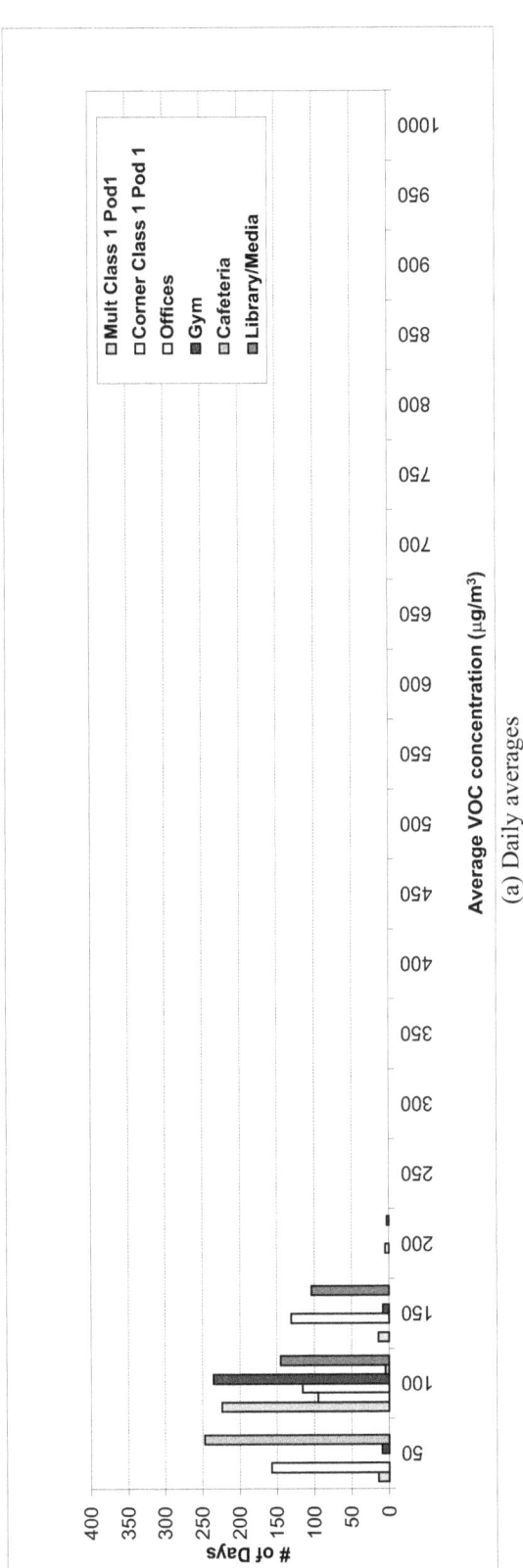

(a) Daily averages

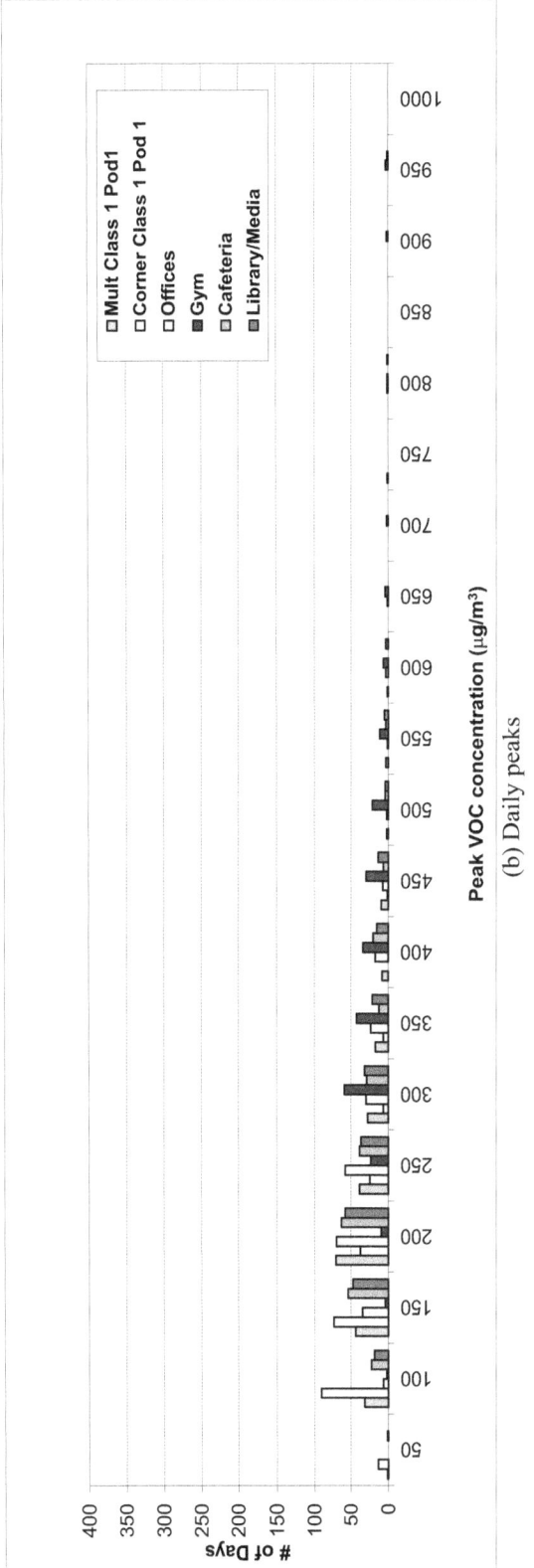

(b) Daily peaks

Figure 29 Frequency distribution of simulated VOC concentration for Primary School

43

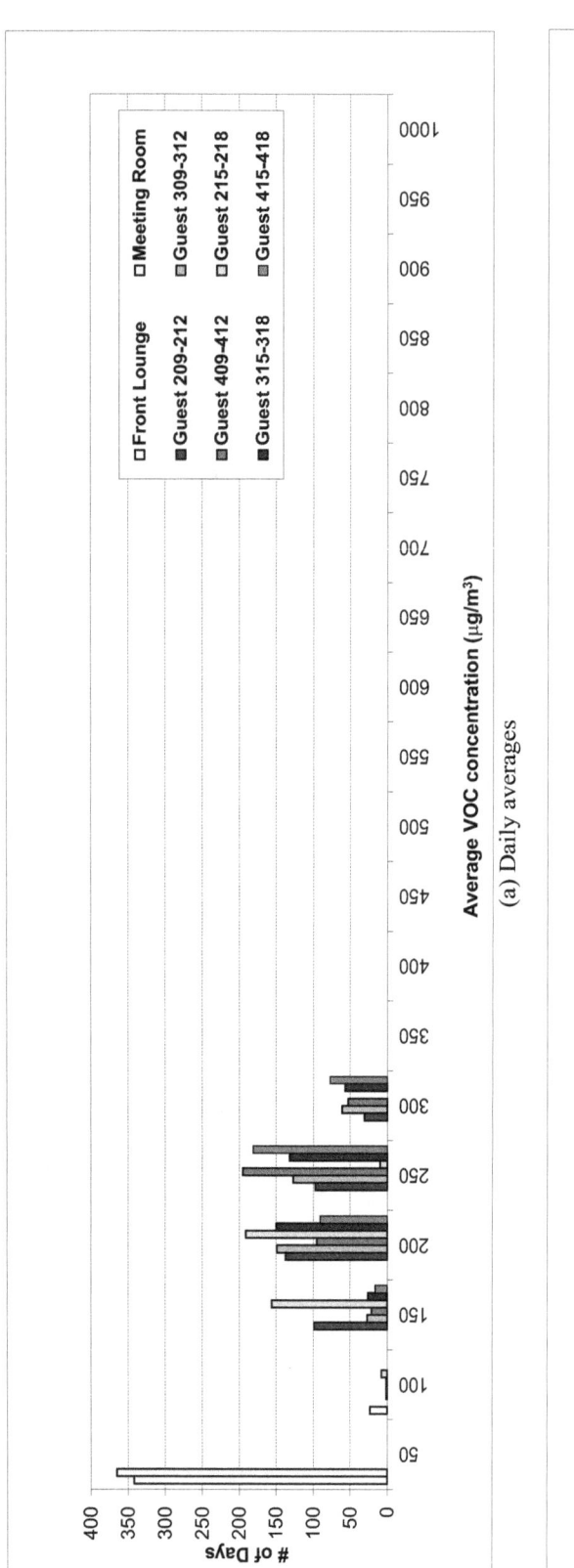

(a) Daily averages

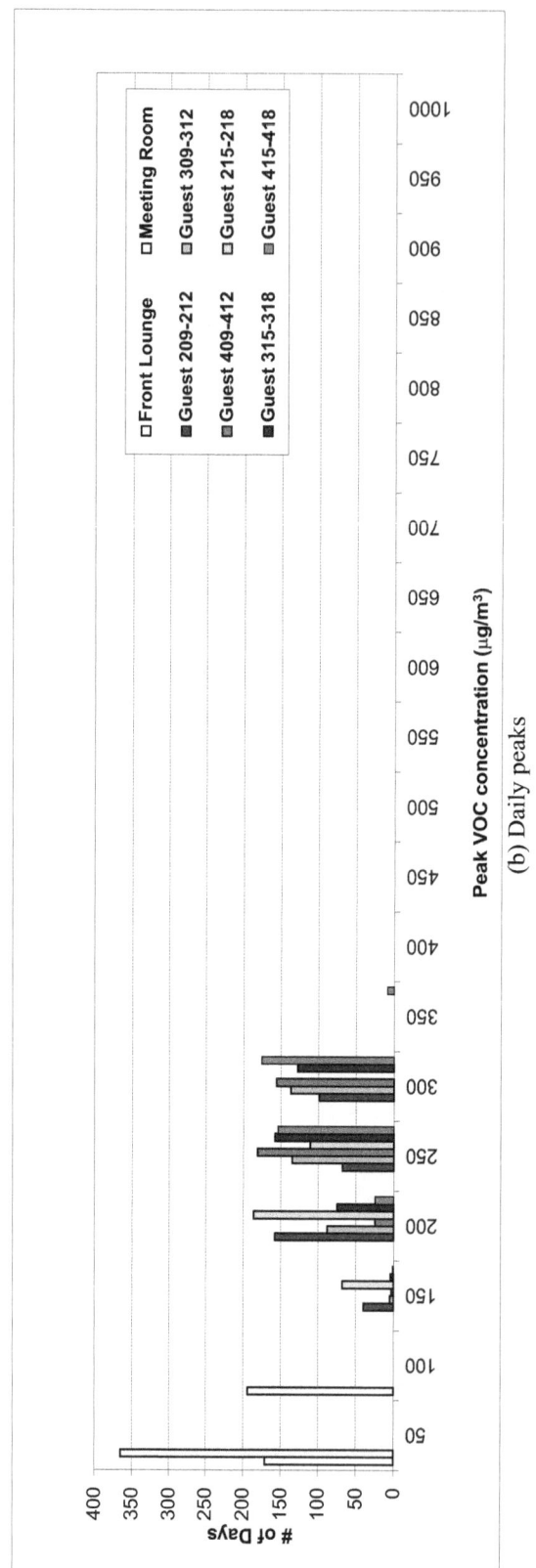

(b) Daily peaks

Figure 30 Frequency distribution of simulated VOC concentration for Small Hotel

44

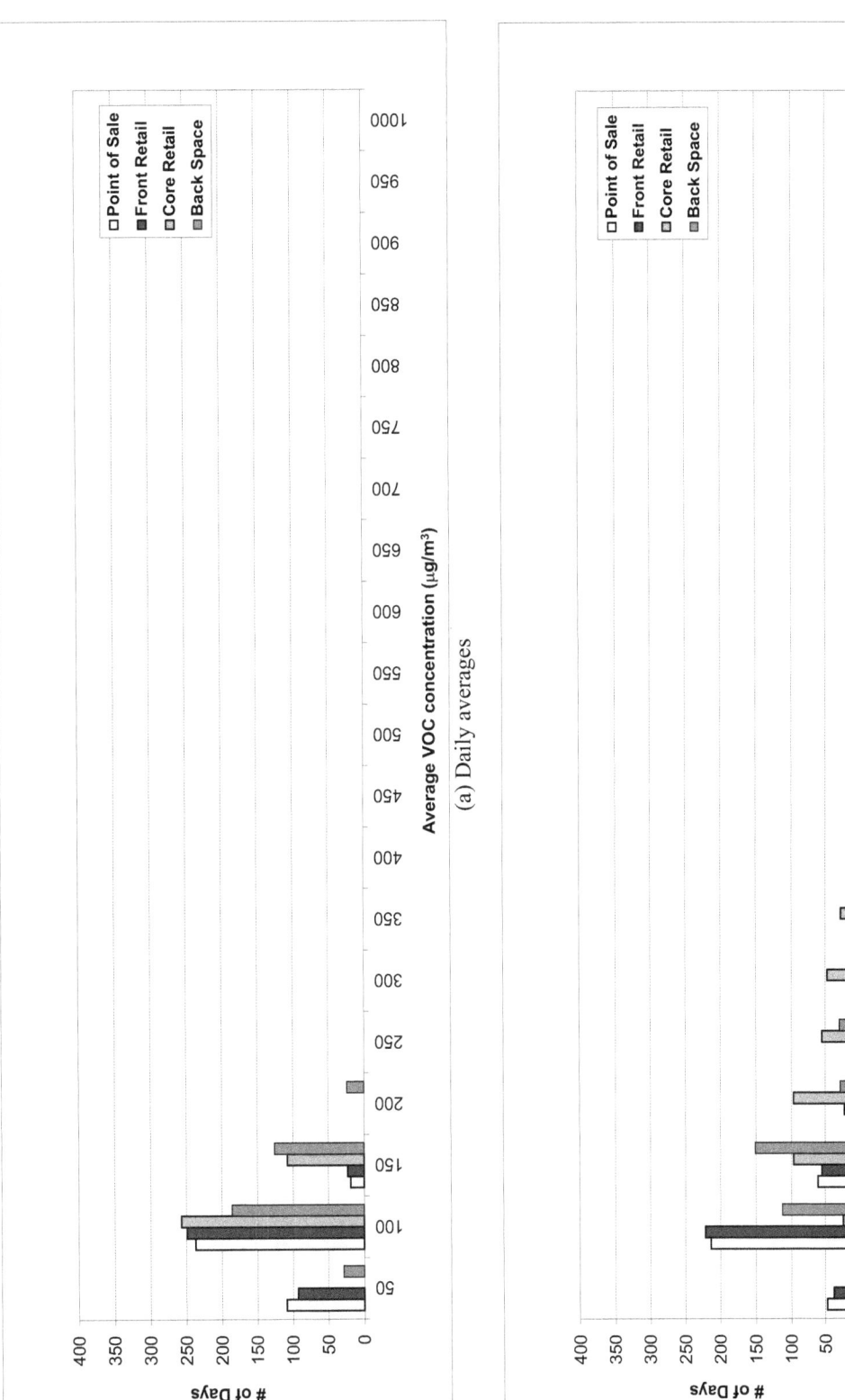

(a) Daily averages

(b) Daily peaks

Figure 31 Frequency distribution of simulated VOC concentration for Stand-Alone Retail

45

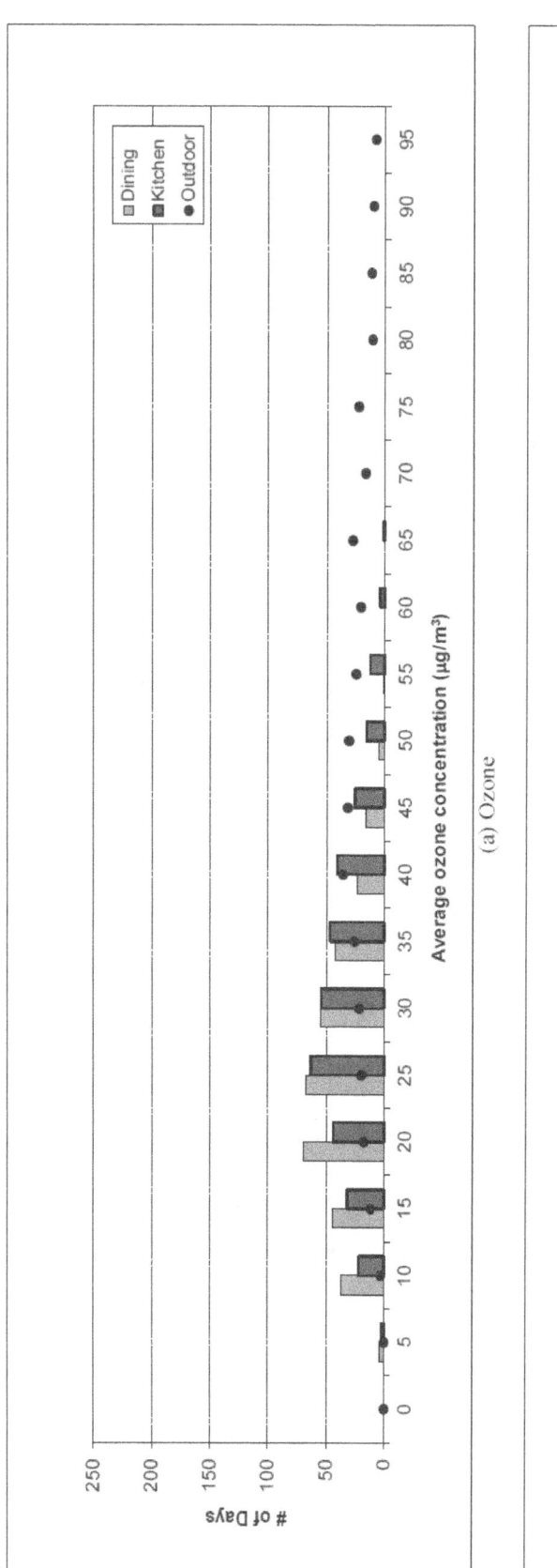

(a) Ozone

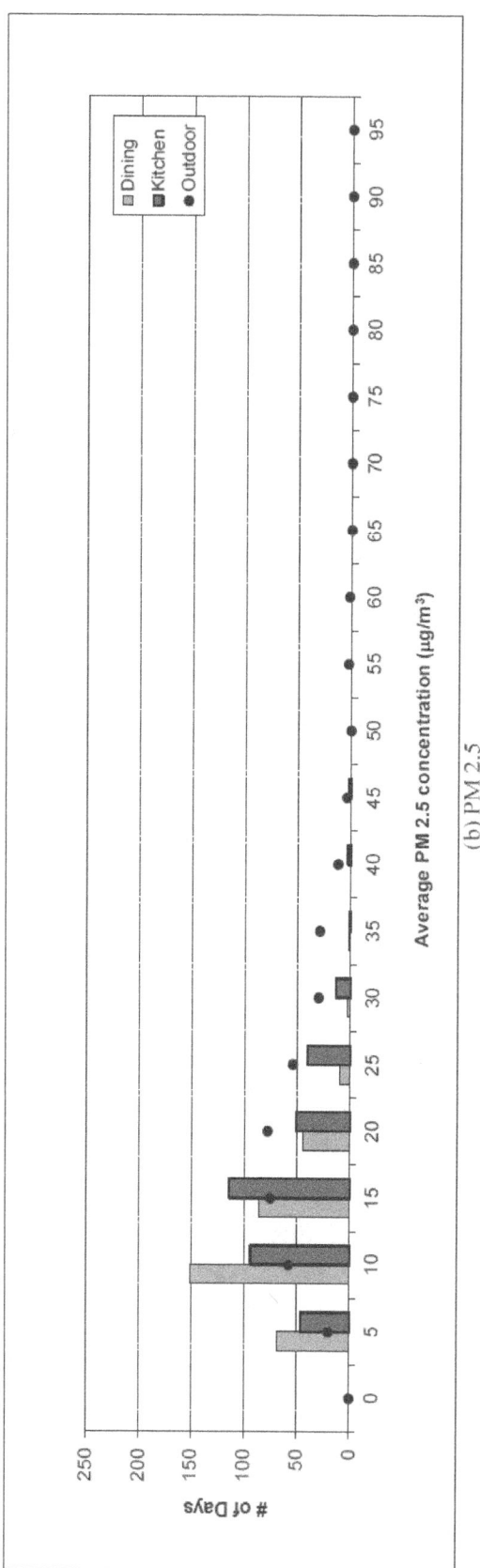

(b) PM 2.5

Figure 32 Frequency distributions of simulated ozone and PM2.5 daily average concentrations for Full Service Restaurant

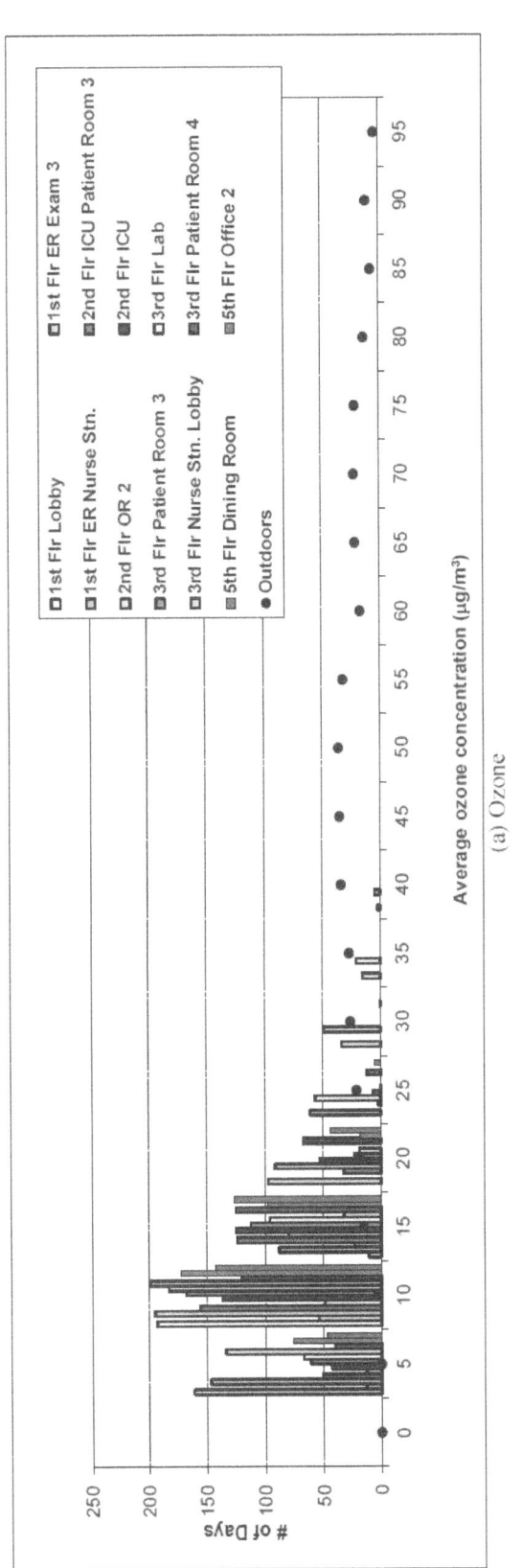

(a) Ozone

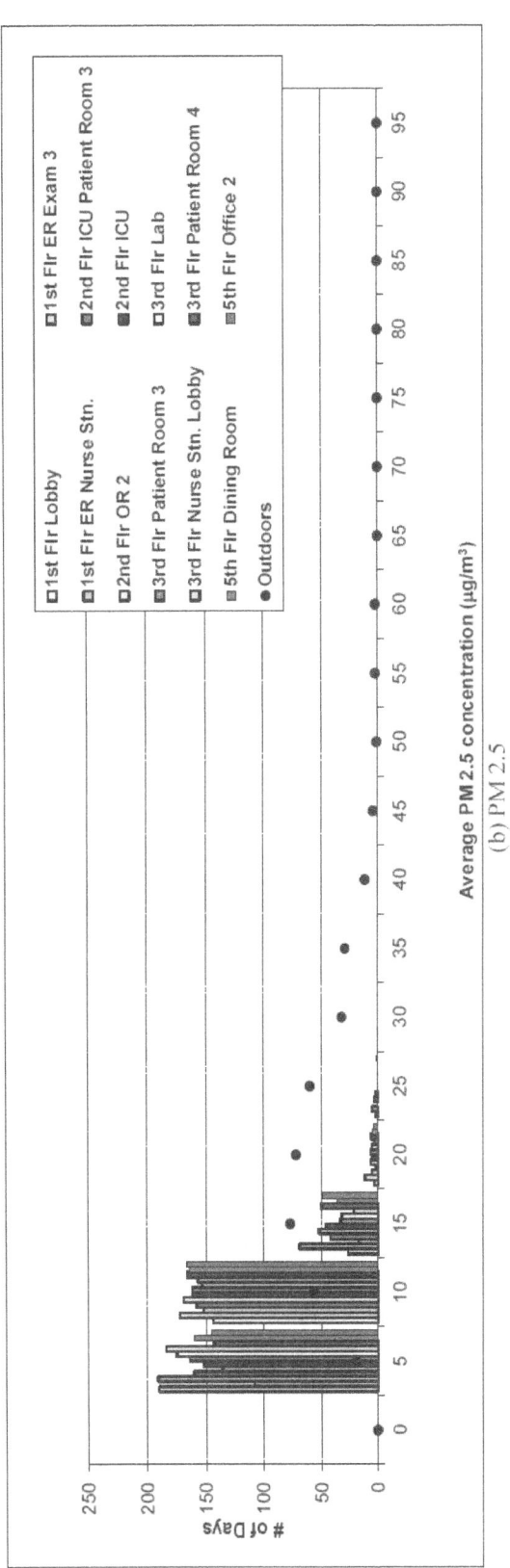

(b) PM2.5

Figure 33 Frequency distributions of simulated ozone and PM2.5 daily average concentrations for Hospital

47

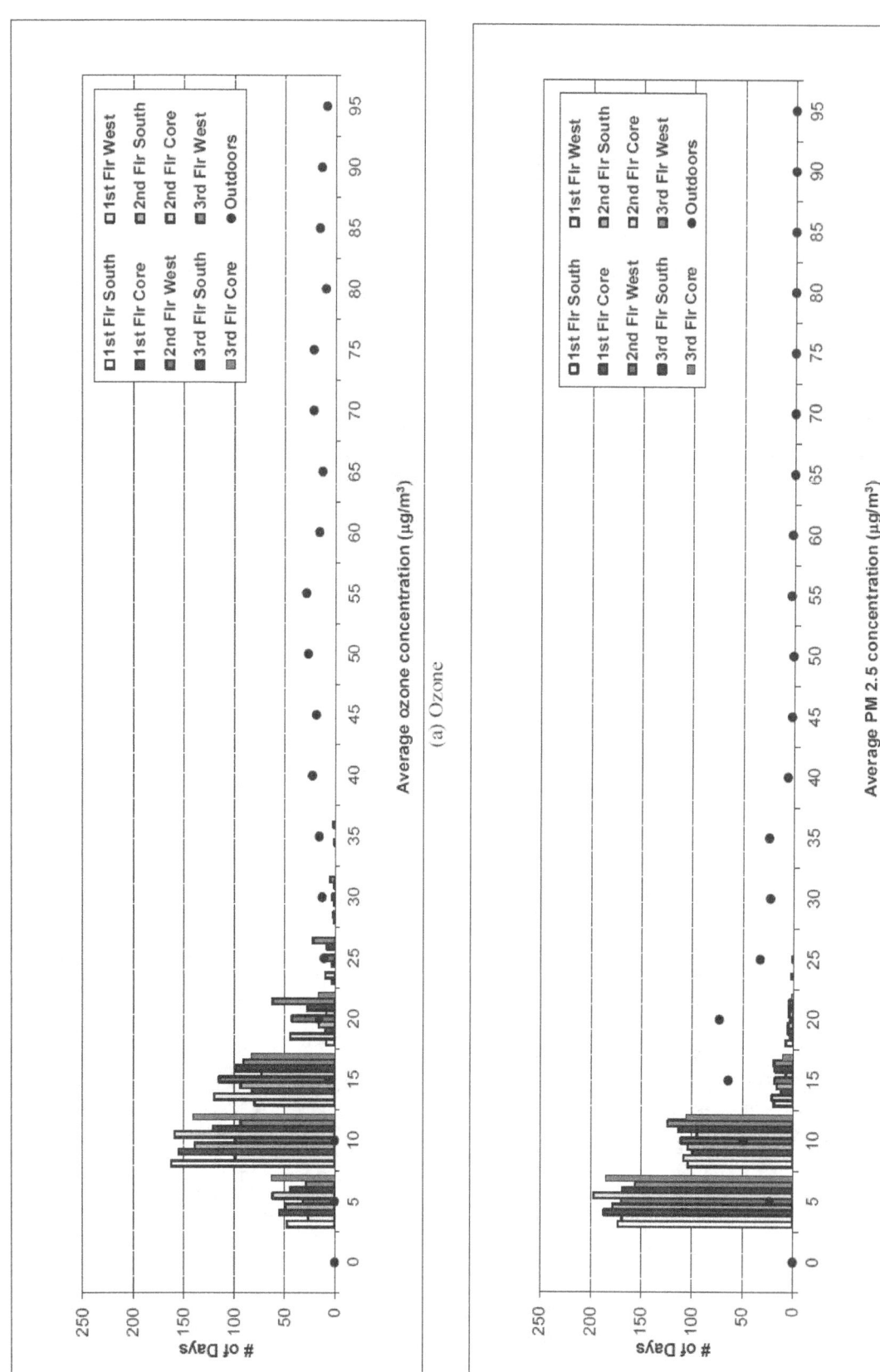

(a) Ozone

(b) PM 2.5

Figure 34 Frequency distributions of simulated ozone and PM2.5 daily average concentrations for Medium Office

48

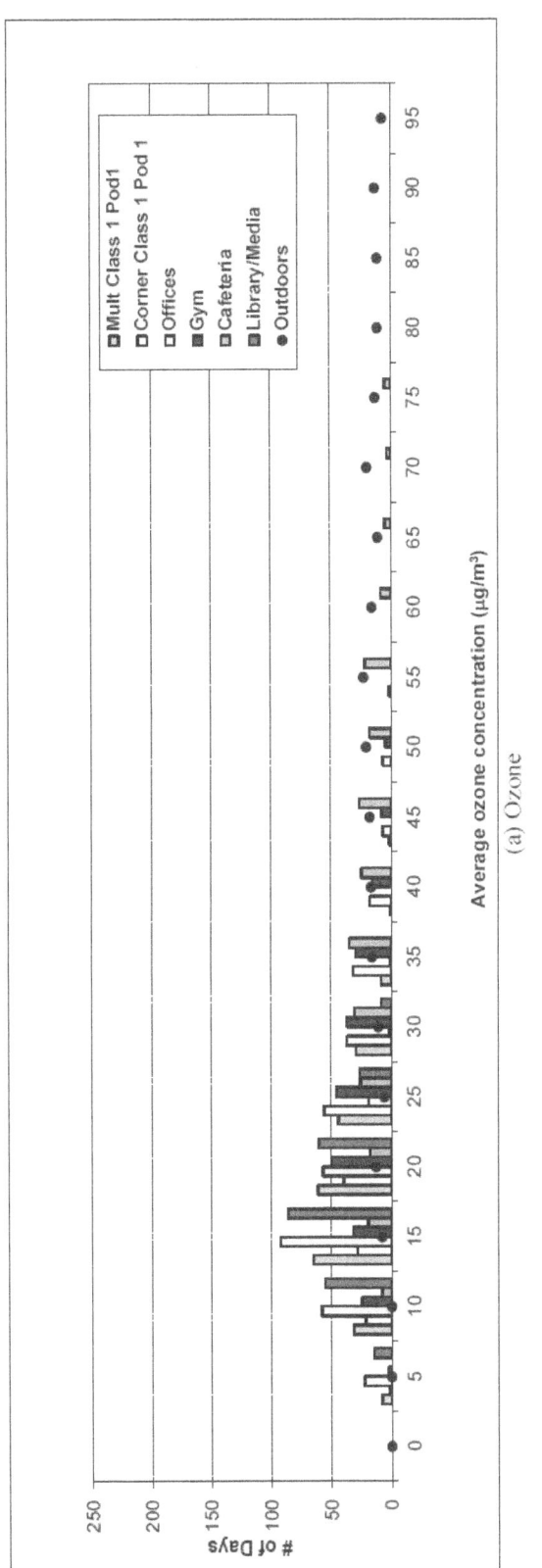

(a) Ozone

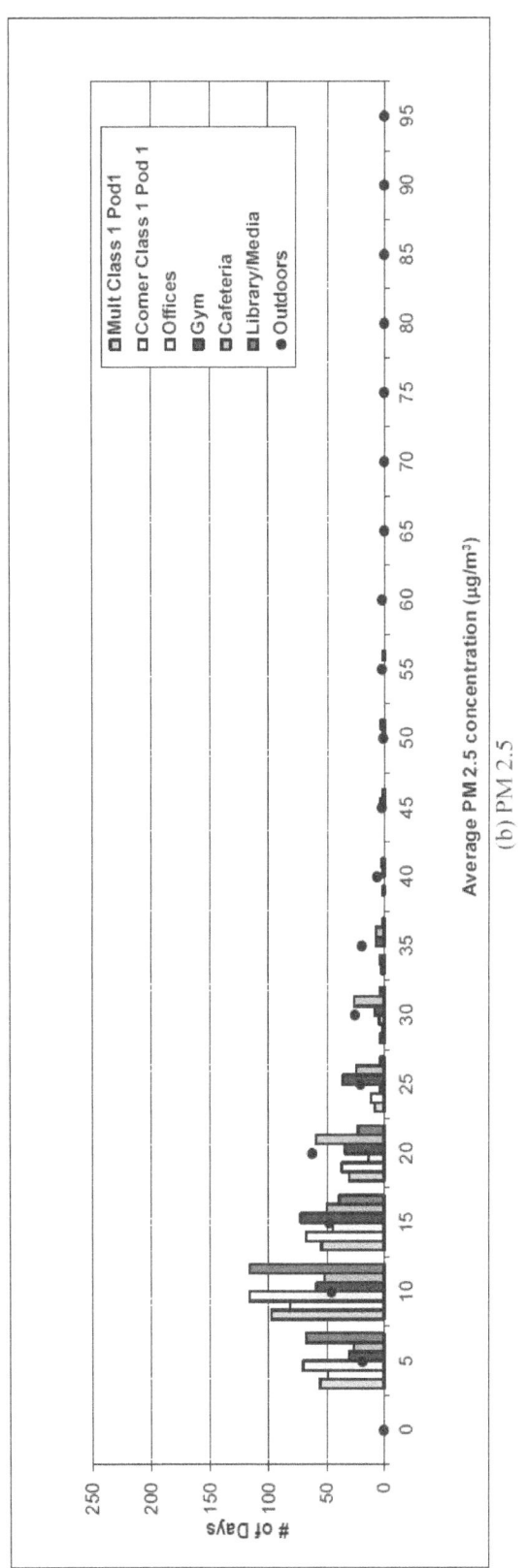

(b) PM 2.5

Figure 35 Frequency distributions of simulated ozone and PM2.5 daily average concentrations for Primary School

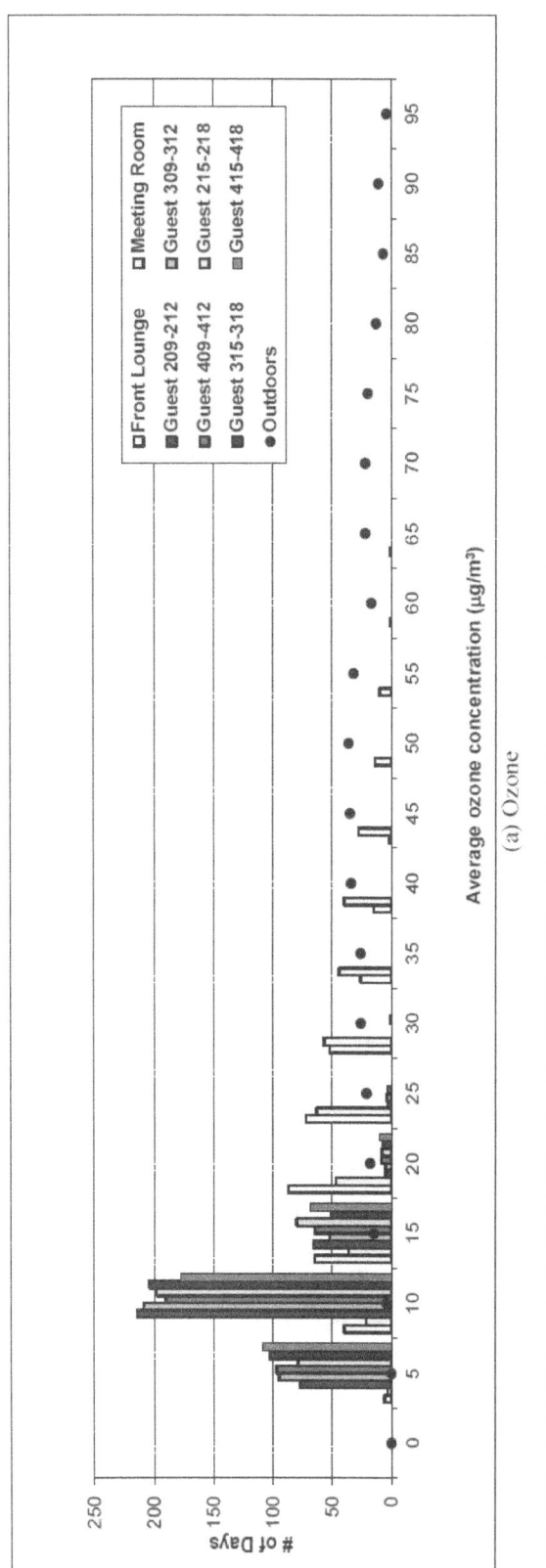

(a) Ozone

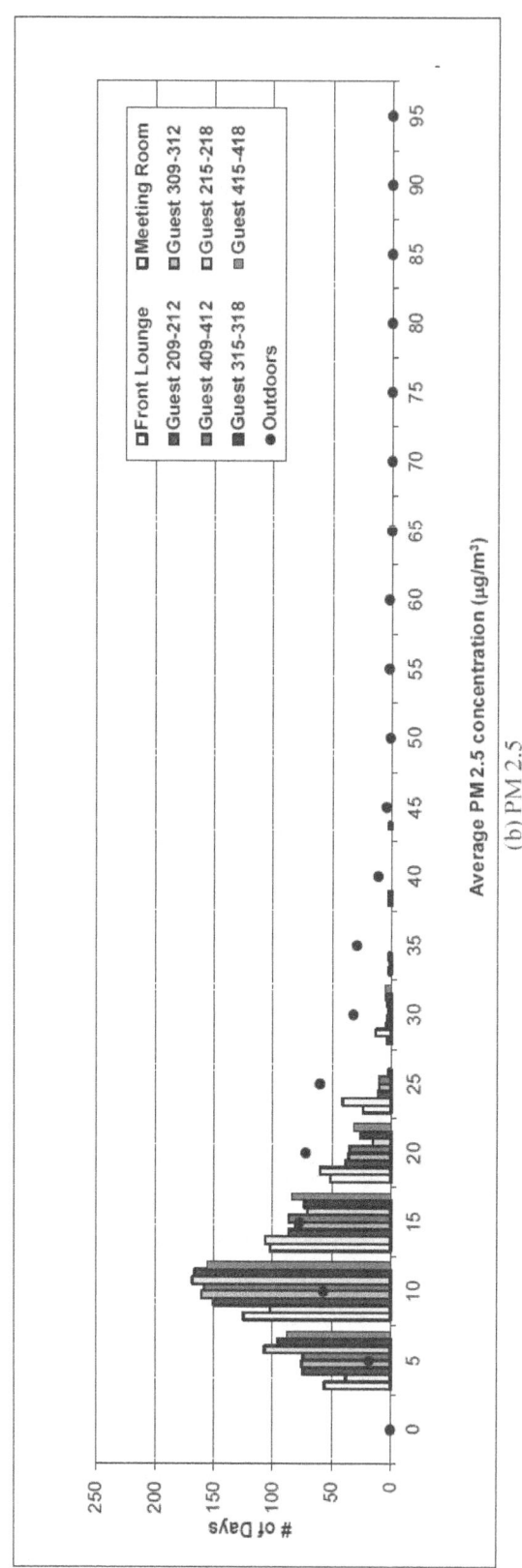

(b) PM 2.5

Figure 36 Frequency distributions of simulated ozone and PM2.5 daily average concentrations for Small Hotel

50

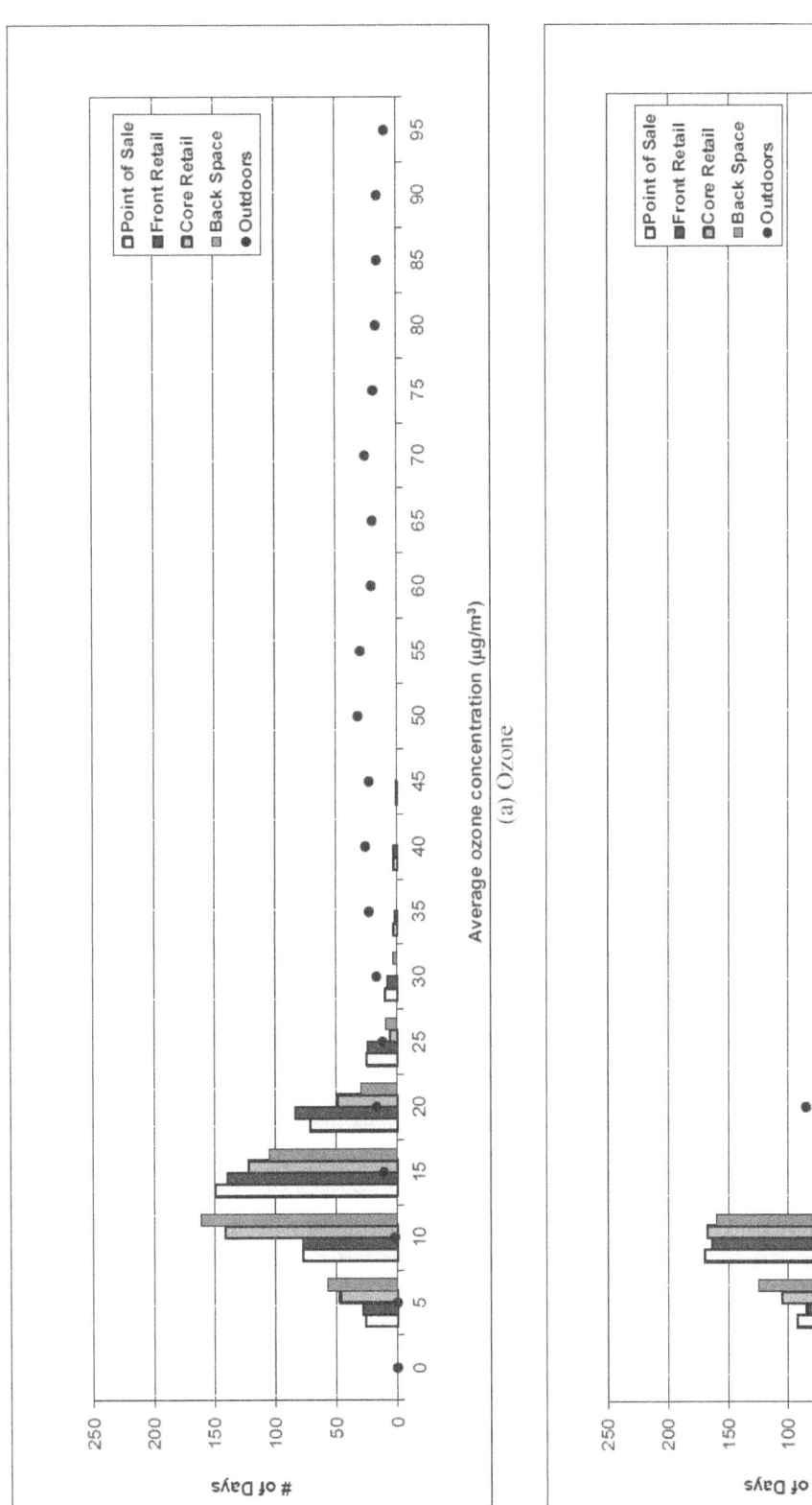

(a) Ozone

(b) PM 2.5

Figure 37 Frequency distributions of simulated ozone and PM2.5 daily average concentrations for Stand-Alone Retail

51

6. DISCUSSION

The development of the CONTAM models of the reference buildings and their application to airflow and contaminant transport analyses described in this report will support future studies of ventilation and IAQ. However, their development presented a number of challenges and other issues that merit discussion and will hopefully be addressed by additional studies in the future.

6.1. Building models for airflow and energy analyses

While the development of CONTAM models of the DOE reference buildings furthers the ability to conduct simultaneous energy, airflow and contaminant transport simulations, it also reveals a number of challenges in doing so. One key issue is that building models developed for performing airflow and IAQ analyses employ different building representations and require different data than those used for energy analyses. CONTAM, and other multizone airflow and indoor air quality (IAQ) models, consider buildings as networks of interconnected zones. Airflow rates are then calculated based on physical relationships between flow and pressure analogous to the relationship between heat transfer and temperature differences in energy models. Thus, it is important that multizone building airflow models capture the geometry of the whole building, its weather exposure, interzone leakages, and ventilation system airflows. In contrast, building models for energy analysis are focused on accounting for thermal loads of different building zones, system efficiencies in meeting these loads, and selecting equipment types and sizes. As discussed below, the zones used in energy calculations are therefore based on the similarity and differences between their thermal loads, while the zones used in airflow modeling are based on pressure relationships and differences in ventilation airflows.

For example, zones with small or insignificant thermal loads, such as stairwells, elevator shafts, and restrooms, are often not included in energy models. However, stairwells and elevators shafts need to be included in multizone models because they have a major effect on building pressure relationships and constitute important airflow paths between zones, particularly in taller buildings. In addition, zones, even floors, with similar thermal loads are often "multiplied" in energy models to simplify model inputs. Such spaces are input once and the resulting loads are multiplied without modeling the individual interior spaces. In multizone models, however, each zone and floor must be modeled separately to account for horizontal air movement and contaminant transport and for the effects of elevation on airflow and contaminant transport.

In addition to different approaches to building zoning, another key difference between airflow and energy models is how they manage airflow balances and interzone airflow. EnergyPlus generally maintains a balance between the ventilation flows into (supply) and out of (return and exhaust) each zone. Interzone airflows are sometimes input, but a net airflow balance between entering and leaving air is maintained for each zone. Infiltration airflows are not part of this balance but are considered only as they impact the thermal loads of the zone. In contrast, CONTAM and other multizone airflow models use the mass balance of air for each zone to determine the amount of infiltration and exfiltration for each zone to the outdoors and/or adjacent zones. Therefore, the system flows are an input while temperature differences and wind pressures serve as boundary conditions, which in conjunction with leakage values of the zone boundaries are used determine these infiltration and exfiltration flows.

For example, equal airflows are specified between dining and kitchen zones in the EnergyPlus models of the restaurants. The value of this interzone airflow is based on minimum ventilation requirements for these zones, but does not reflect the actual physics driving the airflow. In contrast, CONTAM and other airflow models actually calculate this airflow based on the interzone pressure difference across the openings between the zones.

Another difference is the manner in which energy and airflow models manage the common practice in commercial building design of maintaining an excess of supply over return airflow to reduce envelope infiltration. As noted above, energy models maintain a net balance between incoming and outgoing ventilation airflows, so an excess of supply air is not reflected in these models. Airflow models on the other hand need to consider these ventilation airflow differences, as they are key in determining building pressure relationships. It's worth noting that while the EnergyPlus models included kitchens with their excess exhaust flow, that flow did not impact the building airflow dynamics. CONTAM, on the other hand, included both kitchen and restroom exhausts and calculated their impacts on building pressures and airflows. Spaces with local exhaust flows must also be included as separate zones to properly account for their effects on building pressures.

The most important differences in how multizone and energy building models handle infiltration is evident in the differences between the building infiltration rates reported in Section 4. The infiltration rates in the EnergyPlus models are input assuming that the indoor-outdoor pressure differences are always negative 4 Pa, i.e., air enters the buildings at all points on the exterior envelope. The airflow results reported in Section 4 demonstrate that the constant infiltration rates assumed in EnergyPlus do not reflect the physical dependence of infiltration on weather conditions. The largest discrepancies occur for larger differences between indoor and outdoor temperatures and for higher wind speeds, resulting in substantial under-prediction of the infiltration rates and the resultant energy impact for the EnergyPlus (tight) models. The infiltration rates and sensible loads calculated using CONTAM were as much as six to nine times greater than the EnergyPlus (tight) results but only 30 % to two times greater than the EnergyPlus (CONTAM-equiv.) results. Thus, the selection of an infiltration or envelope leakage rate to be used in building energy simulation needs to be carefully considered since it can significantly impact the predicted airflow and energy use. However, given the very limited data on building envelope leakage, selecting these values is a significant challenge for both energy analysis and airflow simulation. Nevertheless, assuming constant infiltration airflow rates in energy simulation, no matter the value, cannot capture the important effects of weather on infiltration. Thus, a more accurate treatment of infiltration should be applied to EnergyPlus models and building energy simulation in general. As an alternative to performing multizone analysis using tools like CONTAM, simple empirical relationships between infiltration rates, airtightness, system operation and weather can be developed for use in energy simulation. The challenge is to develop such algorithms that can provide sufficiently accurate infiltration rates for various building sizes, designs and shapes and under various weather conditions.

6.2. Limitations of study

The CONTAM models of the reference buildings provide important tools to evaluate the ventilation and IAQ performance of various building and system design options in conjunction with EnergyPlus analyses. However, there are a number of limitations to the CONTAM models that need to be considered and potentially addressed in the future. As discussed in Section 3.1, to simplify CONTAM modeling, the maximum supply airflow rates calculated by EnergyPlus were used in the CONTAM models. Therefore, variable-air volume (VAV) system effects were not included. Also, the CONTAM simulations maintained a constant indoor temperature and used the minimum amount of outdoor ventilation air specified in EnergyPlus for each zone (or HVAC system) with no economizer cycle. Thus, future applications of CONTAM to these models may consider varying supply airflow rates, varying indoor temperatures, and economizer operation.

It is also important to note that coupled airflow-thermal interaction cannot be fully be captured by performing independent airflow and thermal simulations, which is especially important for modeling natural or hybrid ventilation approaches. Current methods of coupling airflow-thermal simulations include ping-pong, onion, or fully-integrated (Ng and Persily 2011). Ping-pong coupling passes airflow and temperature values between two separate simulations at each time step. Onion coupling passes airflow and temperature values between two separate simulations within each time step until convergence is reached. Lastly, fully-integrated coupling simultaneously solves the airflow and energy equations within a single simulation. Fully-integrated coupling is the most computationally intensive of the three coupling methods, but may more accurately capture the airflow-thermal interactions. The most appropriate coupling method depends in part to the degree with which the airflow-thermal problem is coupled. The more highly coupled the interaction, such as in naturally ventilated buildings where large temperature gradients may exist and are important drivers of airflow, the more sophisticated the coupling method will need to be. Other important factors in selecting an appropriate airflow-thermal coupling method include: achievable convergence of airflow and thermal values, and the differences in time scales between the airflow (on the order of minutes or hours) and thermal (on the order of seconds to hours) problem. Wang et al. (2010) developed a fully-integrated coupling method and compared its performance to an onion-coupled simulation for a buoyancy-driven problem. The study found numerical instabilities with the onion-coupled method. The fully-integrated coupling method was not subject to these instabilities and predicted airflow rates and temperatures comparable to the results of a CFD simulation with significantly reduced computational cost. Nevertheless, more development and testing of coupling techniques for a variety of airflow-thermal problems are still needed.

6.3. Presenting IAQ simulation results

In using multizone models for IAQ analysis to conduct IAQ simulations to evaluate different buildings and design approaches, the manner in which to present the simulation results is somewhat challenging. IAQ simulations conducted over a year, or any significant period of time, in a multizone building produces a large amount of data. Running such simulations for different conditions of building operation, source strengths, weather conditions or other parameters quickly multiplies the amount of data produced. Converting these data into understandable formats and drawing useful conclusions is difficult, and there are no standard approaches for doing do.

One challenging aspect of analyzing IAQ simulation results is the comparison of the predicted concentrations to meaningful reference values, given the lack of such reference or limit values for most indoor contaminants. As discussed in Section 5, two of the simulated contaminant concentrations (ozone and PM 2.5) were compared with the NAAQS outdoor air limits and WHO indoor air guidelines. However, the other contaminants simulated, CO_2 and VOCs, do not have guidelines for comparison, let alone formalized, health-based limits. The same lack of concentration limits is also the case for many other indoor air contaminants. As noted in Persily and Emmerich (2012), the diversity of occupants and contaminants and the lack of guidelines for exposure limits to the numerous contaminants present in buildings means that IAQ cannot not be judged as good or bad in terms of contaminant concentration(s) alone. Unlike energy consumption or thermal comfort, which can be quantified in terms well-defined parameters, the complex interaction of ventilation rates, contaminant sources, interaction and removal mechanisms, and occupant behavior make the evaluation of IAQ extremely challenging.

The need for IAQ metrics has been considered previously but no set of metrics has been accepted. TenBrinke et al. (1998) correlated results of occupant surveys with total VOC levels. Hollick and Sangiovanni (2000) developed a metric that accounted for the effects of human health and comfort on individuals by various contaminants. Sofuoglo and Moschandreas (2003) aggregated the concentrations of eight contaminants and correlated the resulting metric with occupant surveys to determine an Indoor Air Pollution Index (IAPI). Jackson et al. (2011) based their assessment of IAQ on the potential health risk of VOC exposure to occupants. Catalina and Iordache (2012) proposed an index that incorporated energy consumption, visual comfort, acoustics, and air change rate, but did not consider contaminant concentrations. It is not yet possible to say which IAQ metric is most useful based on their limited application and fundamentally because each situation is different. Thus, the development of an IAQ metric, or perhaps multiple metrics, that is widely applicable to a range of buildings is still a major need.

6.4. Future work

The EnergyPlus and CONTAM models of the reference buildings serve as baseline cases, which can be used in future analyses to support the design and implementation of strategies to simultaneously reduce building energy use while maintaining or improving IAQ, such as alternative ventilation approaches, enhanced filtration and contaminant source control. Among these ventilation approaches, heat recovery ventilation can maintain outdoor air ventilation rates while providing "free" heat exchange between warmer air returned to the system and colder air entering from outdoors in the winter. Demand control ventilation can reduce outdoor air ventilation rates during periods of low occupancy, thus reducing the energy required to condition outdoor air. Economizer operation can increase outdoor air ventilation while also reducing the amount of mechanical cooling when weather conditions and outdoor air quality are suitable. Source control and enhanced filtration also have the potential to improve IAQ while potentially reducing energy use if ventilation rates can be decreased. Natural and hybrid (or mixed mode) ventilation also has the potential to simultaneously reduce energy consumption and improve IAQ, however the analysis of these and other approaches can be limited by the inability of current simulation tools to model building airflow and contaminants in a physically reasonable fashion.

Given that this study considered a limited set of contaminants, future work should include additional contaminants, including those that are unique to specific building types. For example, particles from cooking in the Full Service Restaurant and infectious biological agents in the Hospital both merit attention. There are also numerous other contaminants that are known to affect occupant health and comfort, such as carbon monoxide, formaldehyde and individual VOCs (WHO 2005; WHO 2010) that were not simulated in this study. In addition, other simulations may benefit from considering other transport mechanisms, such as particle deposition, absorption/desorption of VOCs, and enhanced filtration.

7. CONCLUSION

The reference buildings created by DOE in EnergyPlus are intended for assessing new technologies and developing energy codes. However, infiltration rates were assumed in the models (not calculated), which may have a substantial impact on predicted energy impacts of some of these technologies. Also, the limited ability of EnergyPlus to model contaminant transport greatly limits the ability to assess the impact of these technologies on IAQ. Based on these limitations, CONTAM models of the 16 reference buildings were created in order to perform airflow and IAQ analyses. Six of the reference buildings, representing each type of occupancy covered by the 15 commercial reference buildings (excluding the Midrise Apartment building), were selected for annual airflow and contaminant simulations.

While the total building outdoor air change rates were similar for the EnergyPlus and CONTAM models, the infiltration rates calculated by CONTAM were two to six times greater than the assumed inputs in the EnergyPlus (tight) models. The infiltration rates calculated with the EnergyPlus (CONTAM-equiv.) models were closer to the CONTAM predictions. More importantly however, the assumed infiltration rates in EnergyPlus did not reflect the impacts of outdoor weather conditions, which were captured by CONTAM. This inability of the EnergyPlus models to account for weather resulted in substantial under-prediction of the energy impact of infiltration rates and associated energy impacts during colder and windier weather conditions.

In all of the selected buildings and zones, the simulated indoor contaminant levels did not exceed limits set by relevant standards and guidelines for most hours of the year. Note that the IAQ simulations in this study only used a limited set of contaminants and relatively constant source strengths. Additional simulations of other contaminants and source strengths, as well as IAQ control technologies, are needed to better understand a range of important IAQ issues.

The EnergyPlus and CONTAM models of the reference buildings serve as baseline cases, which will be useful in future analyses to support the design and implementation of alternative ventilation and IAQ control approaches that can simultaneously reduce building energy use while maintaining if not improving IAQ.

8. REFERENCES

AIA (2006). Architecture 2030: The 2030 Challenge from http://architecture2030.org/2030_challenge/the_2030_challenge.

Allen, R., T. Larson, L. Sheppard, L. Wallace and L. J. S. Liu (2003). Use of Real-Time Light Scattering Data To Estimate the Contribution of Infiltrated and Indoor-Generated Particles to Indoor Air. Environ. Sci. Technol. 37(16): 3484-3492.

ASHRAE (1989). ASHRAE/IES Standard 90.1-1989: Energy Efficient Design of New Buildings Except Low-Rise Residential Buildings. Atlanta: American Society of Heating, Refrigerating and Air-Conditioning Engineers, Inc.

ASHRAE (1999). ASHRAE Standard 62-1999: Ventilation For Acceptable Indoor Air Quality. Atlanta: American Society of Heating, Refrigerating and Air-Conditioning Engineers, Inc.

ASHRAE (2004). ANSI/ASHRAE/IESNA Standard 90.1-2004: Energy Standard for Buildings Except Low-Rise Residential Buildings. Atlanta: American Society of Heating, Refrigerating and Air-Conditioning Engineers, Inc.

ASHRAE (2009). ASHRAE Handbook Fundamentals. Atlanta: American Society of Heating, Refrigerating and Air-Conditioning Engineers, Inc.

ASHRAE (2010a). ANSI/ASHRAE Standard 62.1-2010: Ventilation for Acceptable Indoor Air Quality. Atlanta: American Society of Heating, Refrigerating and Air-Conditioning Engineers, Inc.

ASHRAE (2010b). ANSI/ASHRAE/IESNA Standard 90.1-2010: Energy Standard for Buildings Except Low-Rise Residential Buildings. Atlanta: American Society of Heating, Refrigerating and Air-Conditioning Engineers.

ASHRAE (2010c). ASHRAE Research Strategic Plan 2010-2015: Navigation for a Sustainable Future. Atlanta: American Society of Heating, Refrigerating, and Air-Conditioning Engineers.

ASTM (2010). ASTM E779-10 Standard Test Method for Determining Air Leakage Rate by Fan Pressurization. Philadelphia: American Society of Testing and Materials.

Bekö, G., O. Halás, G. Clausen and C. J. Weschler (2006). Initial studies of oxidation processes on filter surfaces and their impact on perceived air quality. Indoor Air 16(1): 56-64.

Brinke, J. T., S. Selvin, A. T. Hodgson, W. J. Fisk, M. J. Mendell, C. P. Koshland and J. M. Daisey (1998). Development of New Volatile Organic Compound (VOC) Exposure Metrics and their Relationship to "Sick Building Syndrome" Symptoms. Indoor Air 8(3): 140-152.

Catalina, T. and V. Iordache (2012). IEQ assessment on schools in the design stage. Build. Environ. 49(0): 129-140.

Chen, C. and B. Zhao (2011). Review of relationship between indoor and outdoor particles: I/O ratio, infiltration factor and penetration factor. Atmos. Environ. 45(2): 275-288.

Deru, M., K. Field, D. Studer, K. Benne, B. Griffith, P. Torcellini, B. Liu, M. Halverson, D. Winiarski, M. Rosenberg, M. Yazdanian, J. Huang and D. Crawley (2011). U.S. Department of Energy Commercial Reference Building Models of the National Building Stock. NREL/TP-5500-46861. Colorado: National Renewable Energy Laboratory.

DOE (2008). Building Technologies Program 2008 Multi-Year Program Plan from http://www1.eere.energy.gov/buildings/commercial_initiative/goals.html.

DOE. 2010. Building Energy Data Book. Washington: U.S. Department of Energy. 245 pp.

DOE (2011). Commercial Reference Buildings from http://www1.eere.energy.gov/buildings/commercial_initiative/reference_buildings.html.

Emmerich, S. J. and A. K. Persily. (2011). U.S. Commercial Building Airtightness Requirements and Measurements. 32nd Air Infiltration and Ventilation Centre Conference, Belgium.

EPA (2011a). Air Quality System Data Mart from http://www.epa.gov/ttn/airs/aqsdatamart/.

EPA (2011b). National Ambient Air Quality Standards (NAAQS). Washington: U.S. Environmental Protection Agency.

Feustal, H. E. and B. V. Smith (2001). COMIS 3.1 - User's Guide. Berkeley: Lawrence Berkeley National Laboratory.

Hollick, H. H. and J. J. Sangiovanni (2000). A Proposed Indoor Air Quality Metric for Estimation of the Combined Effects of Gaseous Contaminants on Human Health and Comfort from *Air Quality and Comfort in Airliner Cabins, ASTM STP1393*. N. L. Nagda, ed. Philadelphia: American Society for Testing and Materials: 319 pp.

Howard-Reed, C., L. A. Wallace and S. J. Emmerich (2003). Effect of ventilation systems and air filters on decay rates of particles produced by indoor sources in an occupied townhouse. Atmos. Environ. 37(38): 5295-5306.

Jackson, M. C., R. L. Penn, J. R. Aldred, H. I. Zeliger, G. E. Cude, L. M. Neace, J. F. Kuhs and R. L. Corsi. (2011). Comparison of metrics for characterizing the quality of Indoor air. Indoor Air 2011, Austin, TX.

Kowalski, W. J. and W. P. Bahnfleth (2002). MERV filter models for aerobiological applications. Air Media Summer(2002): 13-17.

Kunkel, D. A., E. T. Gall, J. A. Siegel, A. Novoselac, G. C. Morrison and R. L. Corsi (2010). Passive reduction of human exposure to indoor ozone. Build. Environ. 45(2): 445-452.

Liu, D.-L. and W. W. Nazaroff (2001). Modeling pollutant penetration across building envelopes. Atmos. Environ. 35(26): 4451.

Lstiburek, J. W. (2006). Understanding Attic Ventilation. BSD-102. Massachusetts: Building Science Corporation.

Nazaroff, W. W., A. J. Gadgil and C. J. Weschler. (1993). Critique of the use of deposition velocity in modeling indoor air quality. Modeling of Indoor Air Quality and Exposure, Pittsburgh, PA, USA.

Ng, L. C. and A. K. Persily. (2011). Airflow and Indoor Air Quality Analyses Capabilities of Energy Simulation Software. Indoor Air 2011, Austin, TX.

Persily, A. K. and S. J. Emmerich (2012). Indoor Air Quality in Sustainable, Energy Efficient Buildings. HVAC&R Res. 18(1): 1-17.

Persily, A. K., A. Musser, S. J. Emmerich and A. W. Taylor (2003). Simulations of Indoor Air Quality and Ventilation Impacts of Demand Controlled Ventilation in Commercial and Institutional Buildings. NISTIR 7042. Gaithersburg: National Institute of Standards and Technology.

Riley, W. J., T. E. McKone, A. C. K. Lai and W. W. Nazaroff (2002). Indoor Particulate Matter of Outdoor Origin: Importance of Size-Dependent Removal Mechanisms. Environ. Sci. Technol. 36(2): 200-207.

Sofuoglu, S. C. and D. J. Moschandreas (2003). The link between symptoms of off ice building occupants and in-office air pollution: the Indoor Air Pollution Index. Indoor Air 13(4): 332-343.

Swami, M. V. and S. Chandra (1987). Procedures for calculating natural ventilation airflow rates in buildings. FSEC-CR-163-86. Florida: Florida Solar Energy Center.

Thornburg, J., D. S. Ensor, C. E. Rodes, P. A. Lawless, L. E. Sparks and R. B. Mosley (2001). Penetration of Particles into Buildings and Associated Physical Factors. Part I: Model Development and Computer Simulations. Aerosol Sci. Technol. 34(3): 284 - 296.

Tian, L., G. Zhang, Y. Lin, J. Yu, J. Zhou and Q. Zhang (2009). Mathematical model of particle penetration through smooth/rough building envelope leakages. Build. Environ. 44(6): 1144-1149.

Walton, G. N. (1989). AIRNET - A Computer Program for Building Airflow Network Modeling. Gaithersburg: National Institute of Standards and Technology.

Walton, G. N. and W. S. Dols (2005). CONTAM User Guide and Program Documentation. NISTIR 7251. Gaithersburg: National Institute of Standards and Technology.

Wang, L. L., W. S. Dols and S. J. Emmerich (2010). Simultaneous Solutions of Coupled Thermal Airflow Problem for Natural Ventilation in Buildings. HVAC&R Res. Accepted.

Weschler, C. J. (2000). Ozone in Indoor Environments: Concentration and Chemistry. Indoor Air 10(4): 269-288.

Weschler, C. J., H. C. Shields and D. V. Naik (1989). Indoor Ozone Exposures. J. Air Pollut. Control Assoc. 39(12): 1562-1568.

WHO (2005). Air Quality Guidelines: Global Update 2005. Denmark: World Health Organization Europe.

WHO (2010). WHO guidelines for indoor air quality: selected pollutants. Denmark: World Health Organization Europe.

Appendix A Detailed description of CONTAM models of reference buildings

A1. INTRODUCTION

This appendix summarizes the development of multizone network airflow models for the 16 reference buildings developed by the DOE (Deru et al. 2011). The purpose of this appendix is to document the modeling assumptions, decisions, and structure. The models are intended for use in investigating a variety of indoor air quality issues, which will require a working knowledge of how the models were developed.

Multizone Modeling: Background information

Multizone network airflow modeling is a method for calculating building pressures, airflows and contaminant transport. Buildings are represented by a network of "zones" connected by "airflow paths". Zones are discrete volumes of air within which mass is conserved, and that generally have uniform temperature, pressure, and contaminant concentration. Air moves between zones along airflow paths with defined flow rates or pressure-dependent resistance to airflow. Contaminants move through the building with the bulk airflow, and can be generated or removed, and may undergo chemical reactions in CONTAM. The CONTAM software and more information specific to it can be found at http://www.bfrl.nist.gov/IAQanalysis.

Building input specifications

Zoning strategy: The predominant zoning strategy used in the models is one zone per physical room. The zoning generally matches that used in the EnergyPlus models. In some cases, similar rooms are grouped together into a single zone in EnergyPlus in order to simplify thermal calculations. "Multipliers" are used in the EnergyPlus models to indicate that the thermal load for one particular zone is to be applied to several other ones, or simply multiplied. Each zone is served by the same system and has the same occupancy and usage profiles in EnergyPlus. In CONTAM, each zone is also experiences the same wind and stack effects. Examples of zones with multipliers in the EnergyPlus models are classrooms in the schools, guest rooms in the hotels, and examination rooms in the hospital. How these multiplied zones are modeled in CONTAM will be discussed in Section A2.

The three office models are also zoned to match the EnergyPlus models. In each office, each floor consists of five zones – four perimeter and one core zone. This layout is often used in thermal models of buildings with open-office plans. In EnergyPlus, these zones are separated by solid walls, thereby creating temperature differences between the zones. There is no air exchange between the zones in EnergyPlus. In contrast, large openings are modeled between the zones in CONTAM so that air exchange can occur.

In some buildings, zones are added or subdivided to make the models more realistic and useful as multizone models. These changes are noted on the drawings of the individual buildings in Section A2. The changes mostly involve the addition of restrooms, stairwells, and elevator shafts. These are zones whose unique features can significantly influence building airflow, and are thus desirable to include in a multizone network airflow model of a building.

Holiday schedules: Many of the buildings use holidays in their occupancy schedules. For all of the buildings, holidays fall on the following days:

- January 1
- November 11
- December 25
- July 4
- 3rd Monday in January
- 3rd Monday in February
- Last Monday in May
- 1st Monday in September
- 2nd Monday in October
- 4th Thursday in November

The two schools also have different schedules for summer weekdays. Summer is defined as the days from July 1 to end of the day on August 31. Summer weekdays are modeled using a "type 12" day in CONTAM. The CONTAM weather file used with the schools uses the "type 12" day type for all summer weekdays except holidays.

Daylight savings

Daylight savings is implemented from the 2nd Sunday in March to the end of the day on the 1st Sunday in November.

Weather

Steady-state weather is defined in the base models with an outdoor temperature of 20 °C and no wind. The ambient pressure is 1 atm. A wind speed modifier of 0.36 was specified for all exterior leakage paths.

Transient weather: The annual simulations utilized transient weather files. The files were imported into the CONTAM weather file format (.WTH) from a TMY2 weather file for Chicago O-Hare (DOE 2011). Two CONTAM weather files were created – one for the schools, one for the remaining buildings. In both files, the calendar for weekends, holidays, and daylight savings time was set to match the EnergyPlus models:

- January 1 is a Sunday
- Daylight savings is implemented from the 2nd Sunday in March to the end of the day on the 1st Sunday in November
- No weekend holiday rule is used, meaning holidays that fall on a weekend are not observed on the following Monday

The first of the two files, Chicago.wth, is intended for use with all of the buildings except the schools. The second, ChicagoSchools.wth, is identical except that all weekdays between July 1 and August 31 are designated as "type 12" days. A summer weekday HVAC operation is the same as for the rest of the year. A summer weekday occupancy schedule, however, is different than for the rest of the year.

For users who will create CONTAM weather files for other cities, the "type 12" day must be used for summer weekdays in the schools. Otherwise, the building occupancy in CONTAM will not match those in the EnergyPlus models on these days.

A2. BUILDINGS DESCRIPTION

This section summarizes the specific geometry, zones, system flow rates and types, occupants, and schedules used in each of the building models. All flow rates shown are shown for Chicago. In general, minimum ventilation requirements (or "outside air") do not vary between cities or building vintages.

A2.1. Quick Service Restaurant

Table A1 summarizes the zones modeled in CONTAM for the Quick Service Restaurant, their respective sizes, and maximum occupancy.

Table A1 Summary of zones in Quick Service Restaurant

Zone	Area (m^2)	Height (m)	Maximum occupancy
Dining	116	3.05	83.33
Kitchen	100	3.05	6.25
Restroom	16	3.05	0
Attic	232	1.13	0

Geometry:

232.2 m^2 footprint, single-story building with attic. The EnergyPlus model has two zones (not including the Attic) – Dining and Kitchen. In the CONTAM model, a Restroom (shaded in Figure A1) with a footprint of 4 m × 4 m was carved out of the Kitchen.

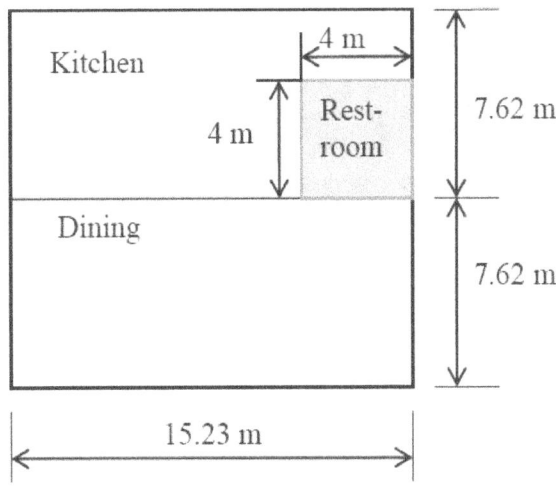

Figure A1 Floor plan of Quick Service Restaurant (height 3.05 m)

The building models for the New and Post-1980 buildings have an attic roof. The Pre-1980 building has a flat roof and no attic zone.

Large interior leakage paths were defined as follows:
- Between the Dining and Kitchen zones, a single large leakage path of 25.7 m² (75 % of the total wall area between the two spaces) is modeled;
- Between the Restroom and Dining zones, a 0.186 m² transfer grille is modeled.

HVAC systems:
For all building vintages, the EnergyPlus model has two packaged constant-volume single-zone systems. In CONTAM, the Dining zone has a constant-volume system. However, the Kitchen is supplied 100 % outside air using a dedicated fan. The supply air, return air, outside air, and exhaust flow rates modeled in CONTAM are listed in Table A2 for all building vintages. The exhaust flow rate for the Restroom was modeled only in CONTAM, not in EnergyPlus.

The EnergyPlus model has a Dining exhaust fan (0.83 m³/s) in addition to the Kitchen exhaust fan (0.72 m³/s). It was included in order to transfer air from the Dining zone to the Kitchen. This is modeled in CONTAM using a large opening between the Dining and Kitchen zones (see above), and one exhaust fan in the Kitchen (1.52 m³/s) and one in the Restroom (0.04 m³/s).

In EnergyPlus, neutral building pressurization is modeled in all zones. Based on discussion with restaurant designers, it is most realistic to model the restaurant with a balanced system in CONTAM as well.

Table A2 Summary of HVAC system flow rates (m³/s) in Quick Service Restaurant

Zone	New		Post-1980		Pre-1980		Outside air (m³/s)	Exhaust air (m³/s)
	Supply	Return	Supply	Return	Supply	Return		
Dining	1.20	0.37	1.20	0.37	1.51	0.68	0.83	0
Kitchen	N/A	N/A	N/A	N/A	N/A	N/A	0.05	1.52
Restroom	N/A	N/A	N/A	N/A	N/A	N/A	0	0.04

Schedules:
All HVAC and exhaust fans operate on the following schedule:
- Every day: on from 5:00 a.m. to 1:00 a.m., off otherwise

Outside air is supplied according to this schedule as well.

Occupants:
The peak number of people for each zone is listed in Table A1. Occupants in all building zones are scheduled according to Figure A2 . There is a different occupancy schedule for weekdays and weekends/holidays.

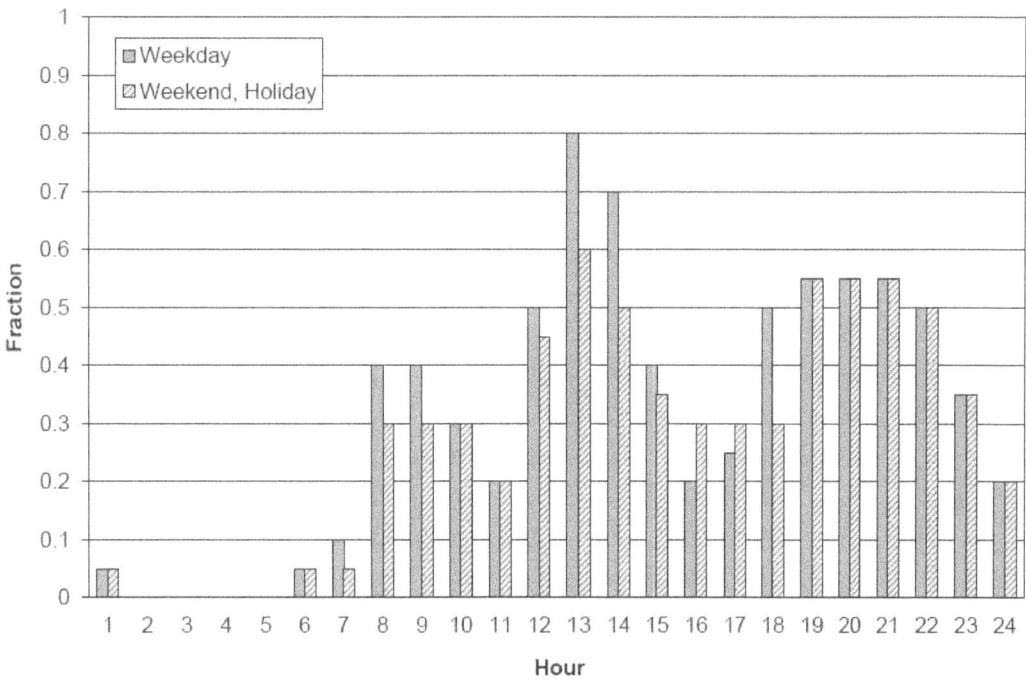

Figure A2 Occupancy schedule for Quick Service Restaurant

A2.2. Full Service Restaurant

Table A3 summarizes the zones modeled in CONTAM for the Full Service Restaurant, their respective sizes, and maximum occupancy.

Table A3 Summary of zones in Full Service Restaurant

Zone	Area (m^2)	Height (m)	Maximum occupancy
Dining	372	3.05	266.77
Kitchen	139	3.05	7.5
Restroom	16	3.05	0
Attic	511	1.68	0

Geometry:

511 m^2 footprint, single-story building with attic. The EnergyPlus model has two zones (not including the Attic) – Dining and Kitchen. In the CONTAM model, a Restroom (shaded in Figure A3) with a footprint of 4 m × 4 m was carved out of the Kitchen.

The building models for new and post-1980 buildings have an attic roof. The Pre-1980 building has a flat roof and no attic zone.

Large interior leakage paths were defined as follows:
- Between the Dining and Kitchen zones, a single large leakage path of 42.57 m^2 (75 % of the total wall area between the two spaces) is modeled;
- Between the Restroom and Dining zones, a 0.186 m^2 transfer grille is modeled.

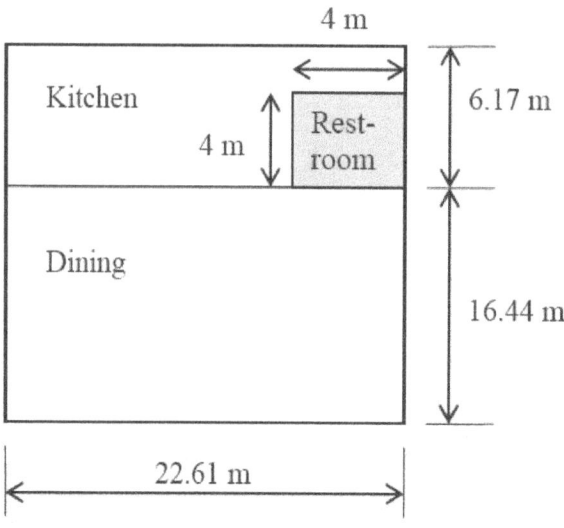

Figure A3 Floor plan of Full Service Restaurant (height 3.05 m)

HVAC systems:

For all building vintages, the EnergyPlus model has two packaged constant-volume single-zone systems. In CONTAM, the Dining zone has a constant-volume system. However, the Kitchen is supplied 100 % outside air using a dedicated fan. The supply air, return air, outside air, and exhaust flow rates modeled in CONTAM are listed in Table A4 for all building vintages. The exhaust flow rate for the Restroom was modeled only in CONTAM, not in EnergyPlus.

The EnergyPlus model has a Dining exhaust fan (1.83 m³/s) in addition to the Kitchen exhaust fan (0.06 m³/s). It was included in order to transfer air from the Dining zone to the Kitchen. This is modeled in CONTAM using a large opening between the Dining and Kitchen zones (see above), and one exhaust fan in the Kitchen (1.85 m³/s) and one in the Restroom (0.04 m³/s).

In EnergyPlus, neutral building pressurization is modeled in all zones. Based on discussion with restaurant designers, it is most realistic to model the restaurant with a balanced system in CONTAM as well.

Schedules:

All HVAC and exhaust fans operate on the following schedule:
- Every day: on from 5:00 a.m. to 1:00 a.m., off otherwise

Outside air is supplied according to this schedule as well.

Table A4 Summary of HVAC system flow rates (m³/s) in Full Service Restaurant

Zone	New		Post-1980		Pre-1980		Outside Air (m³/s)	Exhaust air (m³/s)
	Supply	Return	Supply	Return	Supply	Return		
Dining	2.88	1.72	3.13	1.97	3.64	2.48	2.67	0
Kitchen	N/A	N/A	N/A	N/A	N/A	N/A	0.06	1.85
Restroom	N/A	N/A	N/A	N/A	N/A	N/A	0	0.04

Occupants:

The peak number of people for each zone is listed in Table A3. Occupants in all building zones are scheduled according to Figure A4. There is a different occupancy schedule for weekdays, Saturdays, and Sundays/holidays.

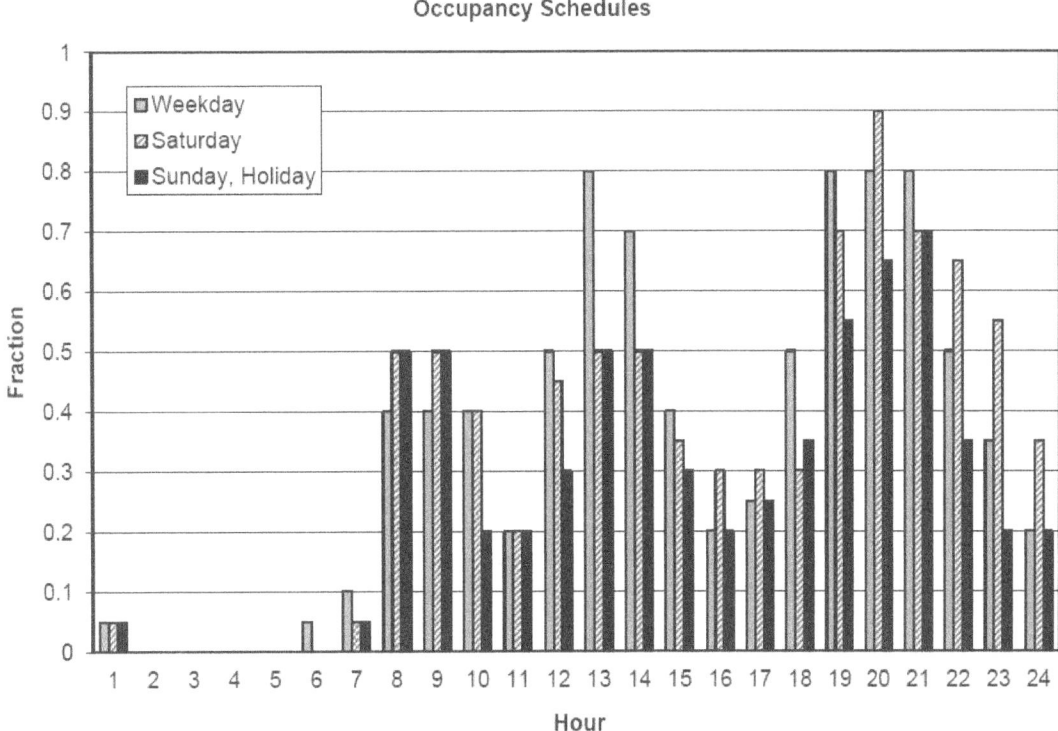

Figure A4 Occupancy schedule for Full Service Restaurant

A2.3.   Small Office

Table A5 summarizes the zones modeled in CONTAM for the Small Office, their respective sizes, and maximum occupancy.

Geometry:

511 m^2 footprint, single-story building with attic. The EnergyPlus model has five zones – four perimeter zones and a core zone. In the CONTAM model, a Restroom (shaded in Figure A5) with a footprint of 4 m × 4 m was carved out of the Core zone.

The building models for new and post-1980 buildings have a vented attic. The Pre-1980 building has insulation above the roof deck, with no attic zone.

Large interior leakage paths were defined as follows:
- Between the perimeter and core zones, a single large leakage path (50 % of the total wall area between the two spaces) is modeled. This is representative of half-height office partitions;
- Between the Restroom and Core zones, a 0.186 m^2 transfer grille is modeled.

Table A5 Summary of zones in Small Office

Zone	Area (m²)	Height (m)	Maximum occupancy
Core	133.66	3.05	8.05
Perimeter South	113.45	3.05	6.11
Perimeter East	67.30	3.05	3.62
Perimeter North	113.45	3.05	6.11
Perimeter West	67.30	3.05	3.62
Attic	511	1.64	0
Restroom	16.0	3.05	0

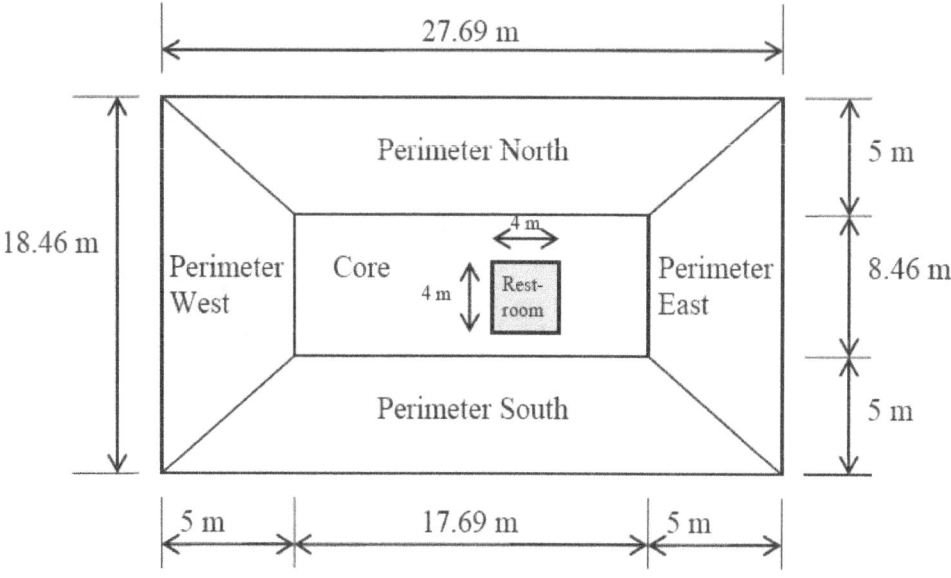

Figure A5 Floor plan of Small Office (height 3.05 m)

HVAC systems:

For all building vintages, the EnergyPlus model has five packaged constant-volume single-zone systems. Similarly, each zone has a constant-volume system in CONTAM. The supply air, return air, outside air, and exhaust flow rates modeled in CONTAM are listed in Table A6 for all building vintages. The exhaust flow rate for the Restroom was modeled only in CONTAM, not in EnergyPlus.

In EnergyPlus, neutral building pressurization is modeled in all zones. To pressurize the building in CONTAM, less air is returned than is supplied to each zone. The return airflow rate is the larger of (a) 90 % of the supply airflow rate and (b) supply airflow rate minus outside air requirement. Return air from the Core zone is reduced by the amount of Restroom exhaust air. The building is neutrally pressurized between 6:00 a.m. and 7:00 a.m. on weekdays and Saturdays when the system operates but no outside air is being supplied.

Table A6 Summary of HVAC system flow rates (m³/s) in Small Office

Zone	New		Post-1980		Pre-1980		Outside Air (m³/s)	Exhaust air (m³/s)
	Supply	Return	Supply	Return	Supply	Return		
Core	0.45	0.37	0.55	0.47	0.58	0.50	0.08	0
Perimeter South	0.36	0.32	0.52	0.47	0.97	0.91	0.06	0
Perimeter East	0.33	0.30	0.37	0.33	0.61	0.57	0.04	0
Perimeter North	0.36	0.32	0.51	0.46	0.96	0.90	0.06	0
Perimeter West	0.36	0.32	0.44	0.40	0.64	0.60	0.04	0
Restroom	N/A	N/A	N/A	N/A	N/A	N/A	0	0.05

Schedules:

All HVAC and exhaust fans operate on the following schedule:
- Weekdays: on from 6:00 a.m. to 10:00 p.m., off otherwise
- Saturday: on from 6:00 a.m. to 6:00 p.m., off otherwise
- Sunday, holidays: off all day

The outside air for the HVAC systems operate on the following schedule:
- Weekdays: on from 7:00 a.m. to 10:00 p.m., off otherwise
- Saturday: on from 7:00 a.m. to 6:00 p.m., off otherwise
- Sunday, holidays: off all day

Occupants:

The peak number of people for each zone is listed in Table A5. Occupants in all building zones are scheduled according to Figure A6. There is a different occupancy schedule for weekdays and Saturdays. Sundays and holidays are unoccupied.

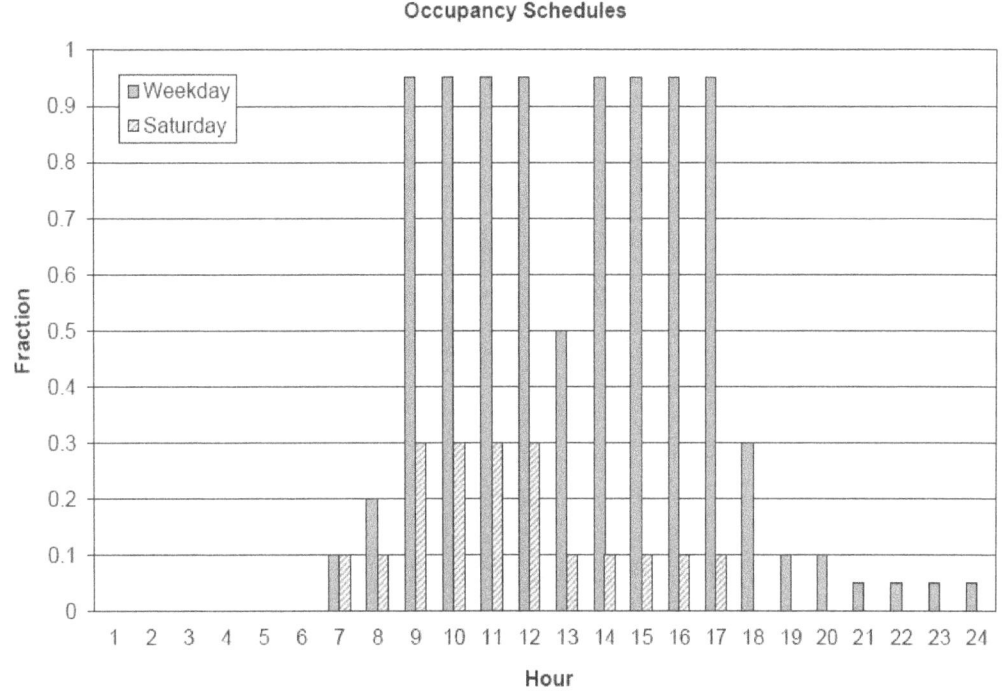

Figure A6 Occupancy schedule for Small Office

A2.4. Medium Office

Table A7 summarizes the zones modeled in CONTAM for the Medium Office, their respective sizes, and maximum occupancy.

Table A7 Summary of zones in Medium Office

Zone	Floor	Area (m²)	Height (m)	Maximum occupancy
Core	1-3	823	2.74	52.93
Perimeter South	1-3	207	2.74	11.16
Perimeter East	1-3	131	2.74	7.06
Perimeter North	1-3	207	2.74	11.16
Perimeter West	1-3	131	2.74	7.06
Restroom	1-3	100	2.74	0
Stairwell	1-3	30	2.74	0
Elevator shaft	1-3	30	2.74	0
Plenum	Abv 1-3	1600	1.26	0
Stair at Plenum	Abv 1-3	30	1.26	0
Elev. at Plenum	Abv 1-3	30	1.26	0

Geometry:

1661 m² footprint, three-story building with flat roof. Total floor area is 4982 m². The EnergyPlus model has five zones per occupied floor – four perimeter zones and a core zone. Floor to ceiling height is 2.74 m and floor to floor height is 4.0 m. The additional height is due to plenums above each floor. Every floor (excluding the plenums) has the same floor plan. All of the plenums have the same floor plans. In the CONTAM model, a Restroom (shaded in Figure A7) with a footprint of 10 m × 10 m was carved out of the Core zone. Also carved out of the Core zone are a 3 m × 10 m Stairwell and a 3 m × 10 m Elevator Shaft (shaded in Figure A7).

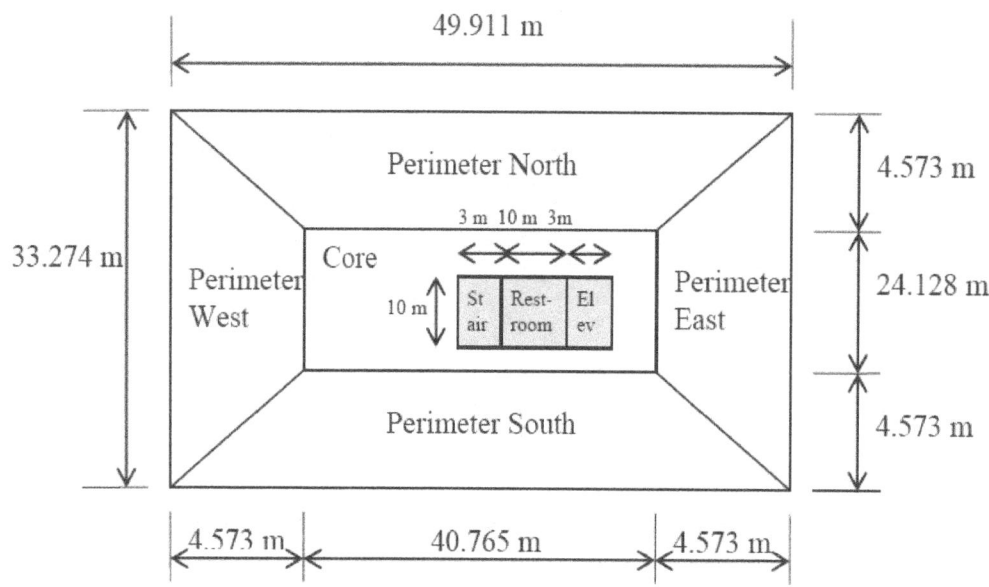

Figure A7 Floor plan of Medium Office (height 2.74 m)

Large interior leakage paths were defined as follows:

- Between the perimeter and core zones, a single large leakage path (50 % of the total wall area between the two spaces) is modeled. This is representative of half-height office partitions;
- Between the Restroom and Core zones, a 0.186 m^2 transfer grille is modeled;
- A leakage path between each occupied zone and the plenum above, equal to 1 % of the floor area of the occupied zone, is modeled to accommodate transfer of return air to the HVAC system through the plenum;
- A stairwell is defined using CONTAM's stair shaft model for closed treads and zero people;
- An elevator shaft is defined using CONTAM's elevator shaft model.

HVAC systems:

For the New and Post-1980 buildings, the EnergyPlus model has three variable air volume (VAV) systems, each serving one floor. The design supply flow rate calculated by EnergyPlus for each VAV system is used as the supply flow rate for each *constant*-volume system modeled in CONTAM for simplicity. The systems modeled in CONTAM are still referred to as "VAV" systems in the body of this text. Varying the supply flow rate can be implemented in CONTAM using controls and/or schedules by users who wish to do so. The supply air, return air, outside air, and exhaust flow rates modeled in CONTAM are listed in Table A8 for the New and Post-1980 buildings. The exhaust flow rates for the Restrooms were modeled only in CONTAM, not in EnergyPlus.

For the New and Post-1980 buildings, the return path for each VAV system in EnergyPlus travels through a plenum above the floor it serves. This is modeled in CONTAM with a large return located in each plenum and a passive opening between the zone and the plenum above that is sized to obtain a maximum velocity of 2 m/s at the grille opening (grille sizes listed in Table A8).

For the Pre-1980 buildings, the EnergyPlus model has 15 constant air volume (CAV) systems, each serving one zone. Similarly, each zone has a constant-volume system in CONTAM. The supply air, return air, outside air, and exhaust flow rates modeled in CONTAM are listed in Table A9 for the Pre-1980 buildings.

In EnergyPlus, neutral building pressurization is modeled in all zones. To pressurize the building in CONTAM, less air is returned than is supplied to each zone. For the New and Post-1980 buildings, the total return plus Restroom exhaust flow rate for each system is set to 90 % of the total supply airflow rate. For the Pre-1980 buildings, the return airflow rate is the larger of (a) 90 % of the supply airflow rate and (b) supply airflow rate minus outside air requirement. Return air from the Core zone is reduced by the amount of Restroom exhaust air. Although the plenums exist in the models of the Pre-1980 building, they are not used as a pathway for airflow, and there are no return grilles in the ceiling. For all building vintages, the building is neutrally pressurized between 6:00 a.m. and 7:00 a.m. on weekdays and Saturdays when the system operates but no outside air is being supplied.

Schedules:

All HVAC and exhaust fans operate on the following schedule:

- Weekdays: on from 6:00 a.m. to 10:00 p.m., off otherwise
- Saturday: on from 6:00 a.m. to 6:00 p.m., off otherwise
- Sunday, holidays: off all day

The outside air for the HVAC systems operate on the following schedule:

- Weekdays: on from 7:00 a.m. to 10:00 p.m., off otherwise
- Saturday: on from 7:00 a.m. to 6:00 p.m., off otherwise
- Sunday, holidays: off all day

Occupants:

The peak number of people for each zone is listed in Table A7. Occupants in all building zones are scheduled according to Figure A8. There is a different occupancy schedule for weekdays and Saturdays. Sundays and holidays are unoccupied.

Table A8 Summary of VAV system flow rates (m³/s) in Medium Office for New and Post-1980 buildings

Zone	Floor	New		Post-1980		Return Grille Size (m²)	Outside Air (m³/s)	Exhaust air (m³/s)
		Supply	Return	Supply	Return			
Core	1	3.21		3.51		1.76	0.57	0
Perimeter South	1	0.68		0.76		0.38	0.12	0
Perimeter East	1	0.83		0.87		0.43	0.15	0
Perimeter North	1	0.57		0.66		0.33	0.10	0
Perimeter West	1	1.04		1.10		0.55	0.18	0
Restroom	1	N/A	N/A	N/A	N/A	N/A	0	0.10
VAV 1 total		**6.32**	**5.59**	**6.90**	**6.11**		**1.12**	
Core	2	3.09		3.47		1.73	0.51	0
Perimeter South	2	0.87		0.97		0.49	0.14	0
Perimeter East	2	0.96		1.02		0.51	0.16	0
Perimeter North	2	0.75		0.85		0.43	0.12	0
Perimeter West	2	1.17		1.23		0.62	0.19	0
Restroom	2	N/A	N/A	N/A	N/A	N/A	0	0.10
VAV 2 total		**6.85**	**6.06**	**7.54**	**6.69**		**1.12**	
Core	3	3.05		3.30		1.65	0.48	0
Perimeter South	3	0.94		0.99		0.50	0.15	0
Perimeter East	3	0.95		0.99		0.50	0.15	0
Perimeter North	3	0.86		0.91		0.46	0.14	0
Perimeter West	3	1.25		1.28		0.64	0.20	0
Restroom	3	N/A	N/A	N/A	N/A	N/A	0	0.10
VAV 3 total		**7.04**	**6.24**	**7.48**	**6.63**		**1.12**	

Table A9 Summary of CAV system flow rates in Medium Office for Pre-1980 building

Zone	Floor	Pre-1980		Outside Air (m³/s)
		Supply	Return	
Core	1	3.53	3.08	0.66
Perimeter South	1	1.05	0.95	0.14
Perimeter East	1	0.99	0.92	0.09
Perimeter North	1	1.04	0.94	0.14
Perimeter West	1	1.25	1.18	0.09
Core	2	3.47	3.02	0.66
Perimeter South	2	1.13	1.02	0.14
Perimeter East	2	1.14	1.07	0.09
Perimeter North	2	1.12	1.01	0.14
Perimeter West	2	1.38	1.31	0.09
Core	3	3.61	3.15	0.66
Perimeter South	3	1.52	1.41	0.14
Perimeter East	3	1.1	1.03	0.09
Perimeter North	3	1.52	1.41	0.14
Perimeter West	3	1.47	1.40	0.09

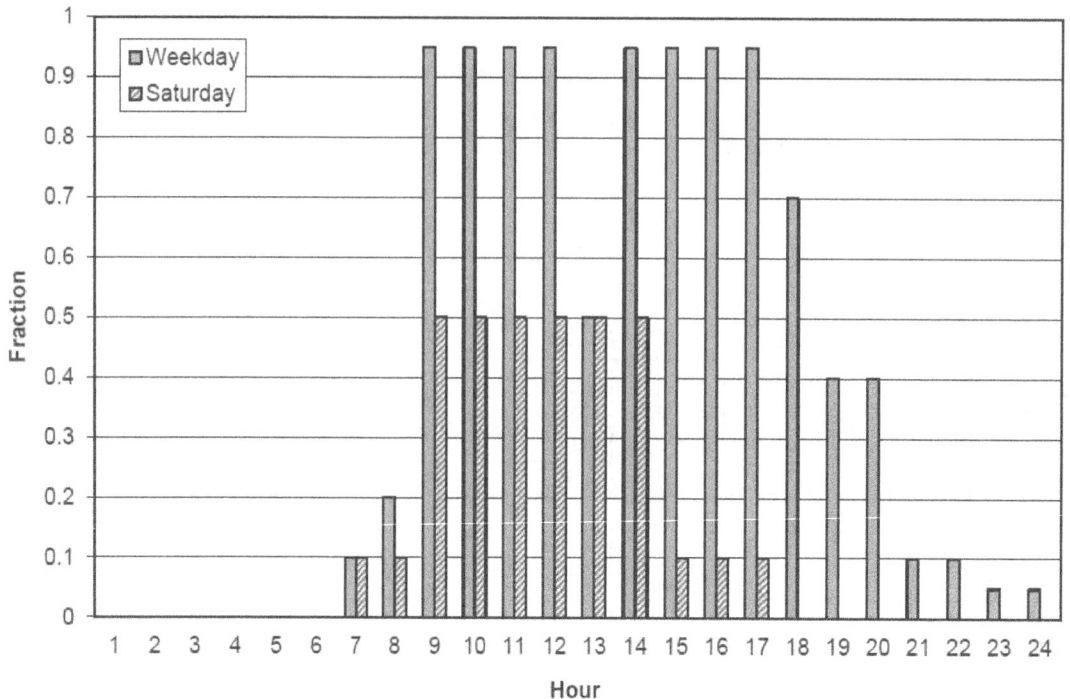

Figure A8 Occupancy schedule for Medium Office

A2.5. Large Office

Table A10 summarizes the zones modeled in CONTAM for the Large Office, their respective sizes, and maximum occupancy.

Geometry:

$3563~m^2$ footprint, 12-story building (plus a basement) with flat roof. Total floor area is $46~320~m^2$ which includes the Basement. The EnergyPlus model has five zones per occupied floor – four perimeter zones and a core zone. Floor to ceiling height is 2.74 m and floor to floor height is 4.0 m. The additional height is due to plenums above each floor. The Basement is a single zone with the same footprint as the upper floors, with a height of 2.44 m. There is no plenum between the basement and the first floor. Every floor (excluding the Basement and plenums) has the same floor plan. All of the plenums have the same floor plans. In the CONTAM model, a Restroom (shaded in Figure A9) with a footprint of 10 m × 10 m was carved out of the Core zone. Also carved out of the Core zone are a 3 m × 10 m Stairwell and a 3 m × 10 m Elevator Shaft (shaded in Figure A9).

Large interior leakage paths were defined as follows:

- Between the perimeter and core zones, a single large leakage path (50 % of the total wall area between the two spaces) is modeled. This is representative of half-height office partitions;
- Between the Restroom and Core zones, a $0.186~m^2$ transfer grille is modeled;

74

- A leakage path between each occupied zone and the plenum above, equal to 1 % of the floor area of the occupied zone, is modeled to accommodate transfer of return air to the HVAC system through the plenum;
- A stairwell is defined using CONTAM's stair shaft model for closed treads and zero people;
- An elevator shaft is defined using CONTAM's elevator shaft model.

Table A10 Summary of zones in Large Office

Zone	Floor	Area (m²)	Height (m)	Maximum occupancy
Basement	B	3563.1	2.44	95.88
Perimeter North	1-12	313.4	2.74	16.87
Perimeter East	1-12	313.4	2.74	16.87
Perimeter South	1-12	202.0	2.74	10.87
Perimeter West	1-12	202.0	2.74	10.87
Core	1-12	2372	2.74	136.29
Restroom	1-12	100	2.74	0
Stairwell	1-12	30	2.74	0
Elevator shaft	1-12	30	2.74	0
Plenum	Abv 1-12	3563.1	1.26	0
Stair at Plenum	Abv 1-12	30	1.26	0
Elev. at Plenum	Abv 1-12	30	1.26	0

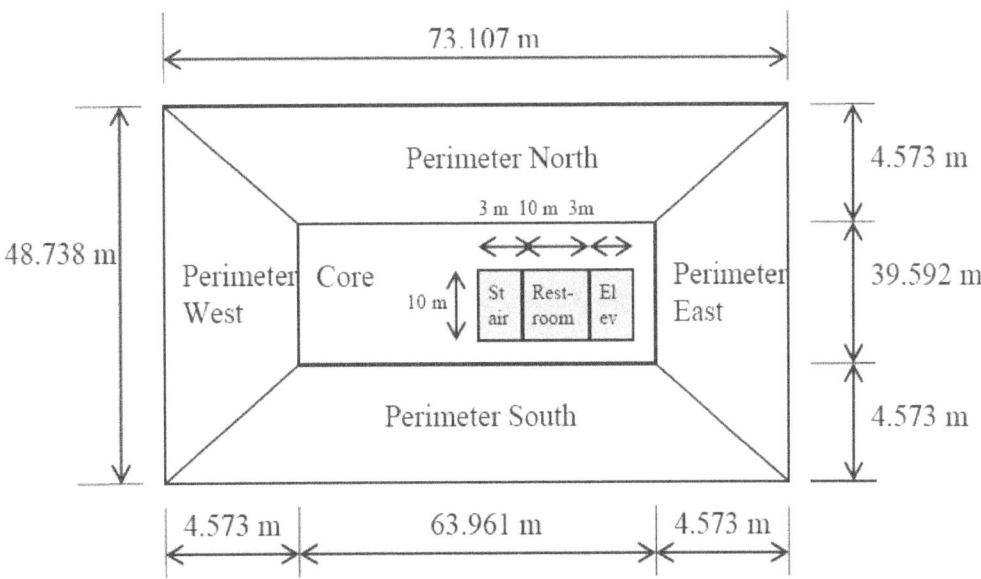

**Figure A9 First floor plan of Large Office (height 2.74 m).
Second through twelfth floors are identical to first floor.**

HVAC systems:
For all building vintages, the EnergyPlus model has four VAV systems. One serves the Basement, one serves the bottom (1st) floor, one serves the "middle" (2nd through 11th) floors, and one serves the top (12th) floor. The design supply flow rate calculated by EnergyPlus for each VAV system is used as the supply flow rate for each *constant*-volume system modeled in CONTAM for simplicity. The systems modeled in CONTAM are still referred to as "VAV" systems in the body of this text. Varying the supply flow rate can be implemented in CONTAM using controls and/or schedules by users who wish to do so. The HVAC system, outside air, and exhaust flow rates modeled in CONTAM are listed in Table A11 for building vintages. The exhaust fans for the Restrooms were modeled only in CONTAM, not in EnergyPlus.

In EnergyPlus, a "middle floor" is multiplied by a factor of 10. Thus, a single VAV system serving the "middle floor" is actually serving 10 floors. In CONTAM, the middle floor VAV system is modeled as 10 separate systems.

The Basement has no plenum, so air is returned back to its system. For the remaining VAV systems, the return path travels through a plenum above that floor. This is modeled in CONTAM with a large return located in each plenum and a passive opening between the zone and the plenum above that is sized to obtain a maximum velocity of 2 m/s at the grille opening.

In EnergyPlus, neutral building pressurization is modeled in all zones. To pressurize the building in CONTAM, less air is returned than is supplied to each zone. For all building vintages, the total return plus Restroom exhaust flow rate for each system (no Restroom exhaust in Basement) is set to 90 % of the total supply airflow rate. The building is neutrally pressurized between 6:00 a.m. and 7:00 a.m. weekdays and Saturdays when the system operates but no outside air is being supplied.

Schedules:
All HVAC and exhaust fans operate on the following schedule:
- Weekdays: on from 6:00 a.m. to 10:00 p.m., off otherwise
- Saturday: on from 6:00 a.m. to 6:00 p.m., off otherwise
- Sunday, holidays: off all day

The outside air for the HVAC systems operate on the following schedule:
- Weekdays: on from 7:00 a.m. to 10:00 p.m., off otherwise
- Saturday: on from 7:00 a.m. to 6:00 p.m., off otherwise
- Sunday, holidays: off all day

Occupants:
The peak number of people for each zone is listed in Table A10. Occupants in all building zones are scheduled according to Figure A8. There is a different occupancy schedule for weekdays and Saturdays. Sundays and holidays are unoccupied.

Table A11 Summary of VAV system flow rates (m³/s) in Large Office

Zone	Floor	New		Post-1980		Pre-1980		Return Grille Size (m²)	Outside Air (m³/s)	Exhaust air (m³/s)
		Supply	Return	Supply	Return					
Basement	B	**7.20**	**6.48**	**9.62**	**8.65**	**8.93**	**8.03**	N/A	**1.20**	**0**
Perimeter North	1	1.26		1.48		1.50		0.75	0.20	0
Perimeter East	1	1.71		1.84		1.93		0.96	0.26	0
Perimeter South	1	1.45		1.66		1.79		0.90	0.23	0
Perimeter West	1	1.99		2.14		2.25		1.13	0.30	0
Core	1	8.84		10.40		10.38		5.20	1.40	0
Restroom	1	N/A	N/A	N/A	N/A	N/A	N/A	N/A	0	0.15
"Bottom" system total		**15.26**	**13.58**	**17.52**	**15.62**	**17.86**	**15.94**		**2.40**	
Perimeter North	2-11	1.30		1.49		1.53		0.76	0.19	0
Perimeter East	2-11	1.79		1.90		1.99		1.00	0.26	0
Perimeter South	2-11	1.53		1.73		1.87		0.93	0.23	0
Perimeter West	2-11	2.12		2.25		2.36		1.18	0.30	0
Core	2-11	9.50		10.92		10.97		5.49	1.41	0
Restroom	2-11	N/A	N/A	N/A	N/A	N/A	N/A	N/A	0	0.15
"Middle" system total		**16.24**	**14.46**	**18.30**	**16.38**	**18.72**	**16.80**		**2.40**	
Perimeter North	12	1.37		1.51		1.63		0.81	0.21	0
Perimeter East	12	1.75		1.83		1.90		0.95	0.26	0
Perimeter South	12	1.55		1.71		1.89		0.95	0.24	0
Perimeter West	12	2.19		2.28		2.45		1.22	0.32	0
Core	12	9.25		10.21		9.99		5.11	1.37	0
Restroom	12	N/A	N/A	N/A	N/A	N/A	N/A	N/A	0	0.15
"Top" system total		**16.11**	**14.35**	**17.55**	**15.64**	**17.85**	**15.94**		**2.40**	

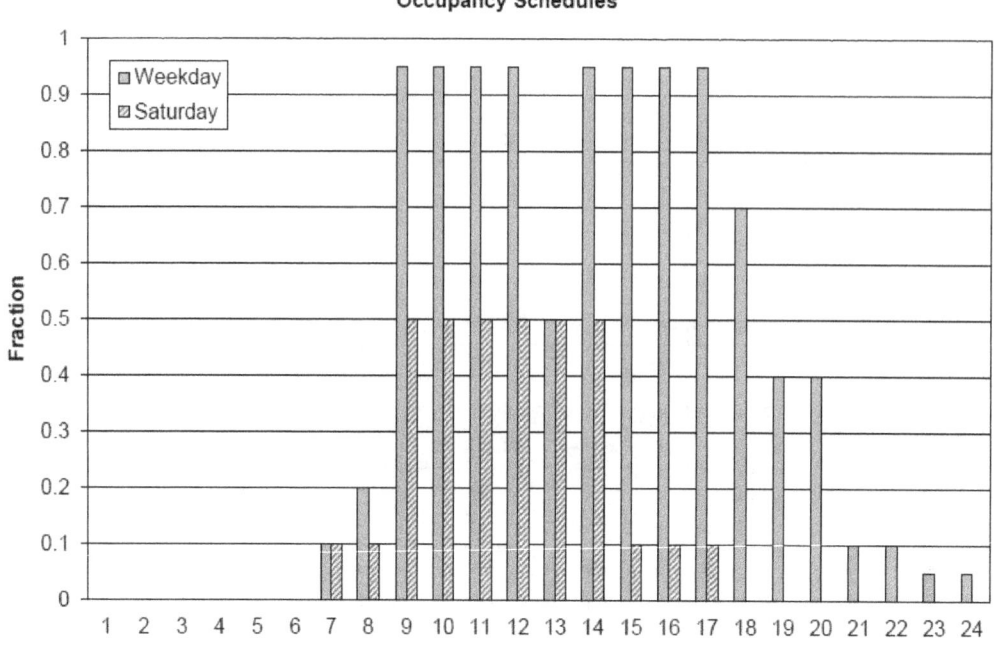

Figure A10 Occupancy schedule for Large Office

A2.6. Primary School

Table A12 summarizes the zones modeled in CONTAM for the Primary School, their respective sizes, and maximum occupancy.

Geometry:

6871 m^2 footprint ("E" shape), one-story building with flat roof. The EnergyPlus model has 25 zones. Most of the large spaces that are typical of schools are modeled as individual zones. Some of the classrooms are grouped together to form a single thermal zone, such as the "Mult" classrooms.

The CONTAM model was altered to create more realistic corridors with reasonable circulation patterns. In EnergyPlus, the Bathroom and Library Media Center blocked access to the Pod 3 corridor. This also prevented access to that section of the building from the Main Corridor. In CONTAM, a 3 m wide path was carved out of the Library Media Center and the Bathroom to connect the Main Corridor to Pod 3 (shaded in Figure A11).

In EnergyPlus, the Mechanical Room extended from the Lobby to the Bathroom, blocking access from the Main Corridor to the Cafeteria, Kitchen, and Gym. In CONTAM, the Mechanical Room was shortened to allow the Main Corridor to provide access to the Cafeteria, Kitchen, and Gym (shaded in Figure A11). These changes make the Main Corridor larger (shaded in Figure A11) and the Library Media Center, Bathroom, and Mechanical Room smaller in the CONTAM model than in the EnergyPlus model. Nevertheless, the occupancies and ventilation rates of the zones were not changed so that the CONTAM model matches the EnergyPlus model.

Table A12 Summary of zones in Primary School

Zone	Area (m^2)	Height (m)	Maximum occupancy
Corner ("Cor") Class 1 Pod 1	99	4.0	24.75
Multiple ("Mult") Class 1 Pod 1	477	4.0	119.25
Corridor ("Corr") Pod 1	192	4.0	19.20
Cor Class 2 Pod 1	99	4.0	24.75
Mult Class 2 Pod 1	477	4.0	119.25
Cor Class 1 Pod 2	99	4.0	24.75
Mult Class 1 Pod 2	477	4.0	119.25
Corr Pod 2	192	4.0	19.20
Cor Class 2 Pod 2	99	4.0	24.75
Mult Class 2 Pod 2	477	4.0	119.25
Cor Class 1 Pod 3	99	4.0	24.75
Mult Class 1 Pod 3	477	4.0	119.25
Corr Pod 3	192	4.0	19.20
Cor Class 2 Pod 3	99	4.0	24.75
Mult Class 2 Pod 3	315	4.0	78.75
Computer Class	162	4.0	48.65
Main Corridor	708	4.0	54.6
Lobby	171	4.0	0
Mechanical Room	156	4.0	0
Bathroom	160	4.0	19.00
Offices	441	4.0	22.05
Library Media Center	363	4.0	91.72
Gym	357	4.0	107.21
Kitchen	168	4.0	25.19
Cafeteria	315	4.0	226.04

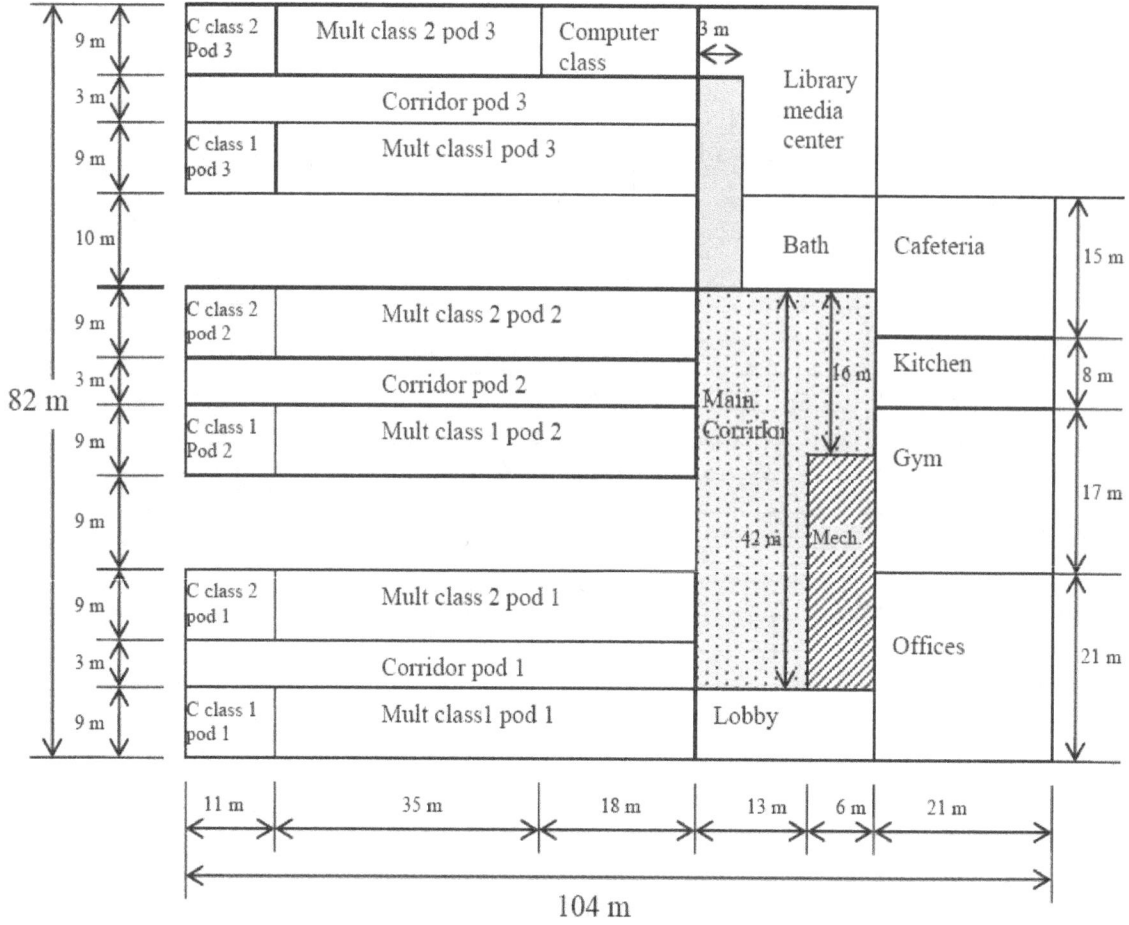

Figure A11 Plan of Primary School (height 4.0 m)

Large interior leakage paths were defined as follows:
- Between the Kitchen and Cafeteria zones, a single large leakage path of 42 m^2 (50 % of the total wall area between the two spaces) is modeled;
- Between each Pod's corridor and the Main Corridor, a 4.0 m^2 open doorway is modeled;
- Between the Lobby and Main Corridor, two 4.0 m^2 open doorways are modeled;
- Between the Bathroom and Main Corridor, a 0.186 m^2 transfer grille is modeled.

HVAC systems:
For all building vintages, the EnergyPlus model has three CAV and four VAV systems. The Kitchen, Gym, and Cafeteria each have a CAV system. The VAV systems are zoned as follows:
- VAV Pod 1: serves the zones in Pod 1
- VAV Pod 2: serves the zones in Pod 2
- VAV Pod 3: serves the zones in Pod 3
- VAV Other: serves the Computer Class, Main Corridor, Lobby, Mechanical Room, Bathroom, Offices, and Library Media Center (7 zones total)

80

The design supply flow rate calculated by EnergyPlus for each VAV system is used as the supply flow rate for each *constant*-volume system modeled in CONTAM for simplicity. The systems modeled in CONTAM are still referred to as "VAV" systems in the body of this text. Varying the supply flow rate can be implemented in CONTAM using controls and/or schedules by users who wish to do so. The supply air, return air, outside air, and exhaust flow rates modeled in CONTAM are listed in Table A13 for all building vintages.

In EnergyPlus, there was a Cafeteria exhaust fan (1.3554 m^3/s) in addition to the Kitchen exhaust fan (0.2016 m^3/s). It was included in order to transfer air from the Cafeteria to the Kitchen. This is modeled in CONTAM using a large opening between the Cafeteria and Kitchen zones (see above), and one larger exhaust fan in the Kitchen (1.557 m^3/s).

In EnergyPlus, neutral building pressurization is modeled in all zones. To pressurize the building in CONTAM, less air is returned than is supplied to each zone. For all building vintages, the return airflow rate is set to 90 % of the supply airflow rate. The return airflow from the Bathroom is reduced by the Bathroom exhaust (0.28 m^3/s). The return airflow from the Kitchen is reduced by 0.20 m^3/s and the return airflow from the Cafeteria is reduced by 1.36 m^3/s to allow for the Kitchen exhaust (1.56 m^3/s).

Schedules:
All HVAC and exhaust fans operate on the following schedule:
- Weekdays: on from 6:00 a.m. to 9:00 p.m., off otherwise
- Weekends and holidays: off all day

Outside air is supplied according to this schedule as well. Summer weekday schedules, as discussed in Section A1, are not used in the HVAC/fan schedules.

Occupants:
The peak number of people for each zone is listed in Table A12. Occupants in all building zones are scheduled according to Figure A12 to Figure A13. Sundays and holidays are unoccupied. There are four different occupant schedules for the building. The occupancy schedules for the Gym and Cafeteria are shown in Figure A12. The occupancy schedule for the Offices is shown in Figure A13. Also shown in Figure A13 is the occupancy schedule for the remaining zones (referred to as "Class" occupancy). All of the schedules in Figure A12 to Figure A13 consider "summer" to be July 1 through August 31. "School year" is the remainder of the year.

Table A13 Summary of HVAC system flow rates (m³/s) in Primary School

Zone	New		Post-1980		Pre-1980		Outside Air (m³/s)	Exhaust air (m³/s)
	Supply	Return	Supply	Return	Supply	Return		
Cor Class 1 Pod 1	0.59	0.53	0.61	0.55	0.75	0.68	0.20	0
Mult Class 1 Pod 1	0.56	0.50	0.58	0.53	0.72	0.65	0.95	0
Corr Pod 1	0.42	0.37	0.50	0.45	0.64	0.58	0.10	0
Cor Class 2 Pod 1	1.45	1.31	1.53	1.38	1.96	1.76	0.20	0
Mult Class 2 Pod 1	1.29	1.16	1.38	1.24	1.69	1.52	0.95	0
VAV-Pod 1 total	**4.30**	**3.87**	**4.61**	**4.15**	**5.77**	**5.19**	**2.40**	
Cor Class 1 Pod 2	0.58	0.52	0.61	0.54	0.75	0.67	0.20	0
Mult Class 1 Pod 2	0.56	0.50	0.59	0.53	0.72	0.65	0.95	0
Corr Pod 2	0.41	0.37	0.50	0.45	0.64	0.58	0.10	0
Cor Class 2 Pod 2	1.41	1.27	1.50	1.35	1.92	1.73	0.20	0
Mult Class 2 Pod 2	1.29	1.16	1.38	1.24	1.69	1.52	0.95	0
VAV-Pod 2 total	**4.25**	**3.82**	**4.57**	**4.11**	**5.72**	**5.14**	**2.40**	
Cor Class 1 Pod 3	0.58	0.52	0.61	0.55	0.75	0.67	0.20	0
Mult Class 1 Pod 3	0.56	0.51	0.59	0.53	0.73	0.66	0.95	0
Corr Pod 3	0.42	0.38	0.50	0.45	0.65	0.58	0.10	0
Cor Class 2 Pod 3	1.42	1.28	1.50	1.35	1.92	1.73	0.20	0
Mult Class 2 Pod 3	0.89	0.80	0.94	0.85	1.15	1.04	0.95	0
VAV-Pod 3 total	**3.87**	**3.48**	**4.14**	**3.73**	**5.19**	**4.67**	**2.40**	
Computer Class	0.50	0.45	0.53	0.48	0.64	0.58	0.39	0
Main Corridor	0.93	0.84	1.10	0.99	1.43	1.29	0.27	0
Lobby	0.39	0.35	0.44	0.39	0.51	0.46	0.09	0
Mechanical Room	0.41	0.37	0.31	0.28	0.41	0.37	0.06	0
Bathroom	0.55	0.22	0.51	0.18	0.66	0.31	0.30	0.28
Offices	1.10	0.99	1.18	1.06	1.51	1.36	0.22	0
Library Media Center	1.10	0.99	1.20	1.08	1.47	1.32	0.73	0
VAV-Other total	**4.99**	**4.21**	**5.28**	**4.47**	**6.63**	**5.69**	**2.07**	
Gym (CAV 2:5)	1.14	1.03	1.07	0.96	1.07	0.96	1.07	0
Kitchen (CAV 1:6)	0.62	0.42	0.60	0.40	0.70	0.50	0.20	1.56
Cafeteria (CAV 2:7)	2.27	0.68	2.27	0.68	2.27	0.68	2.26	0

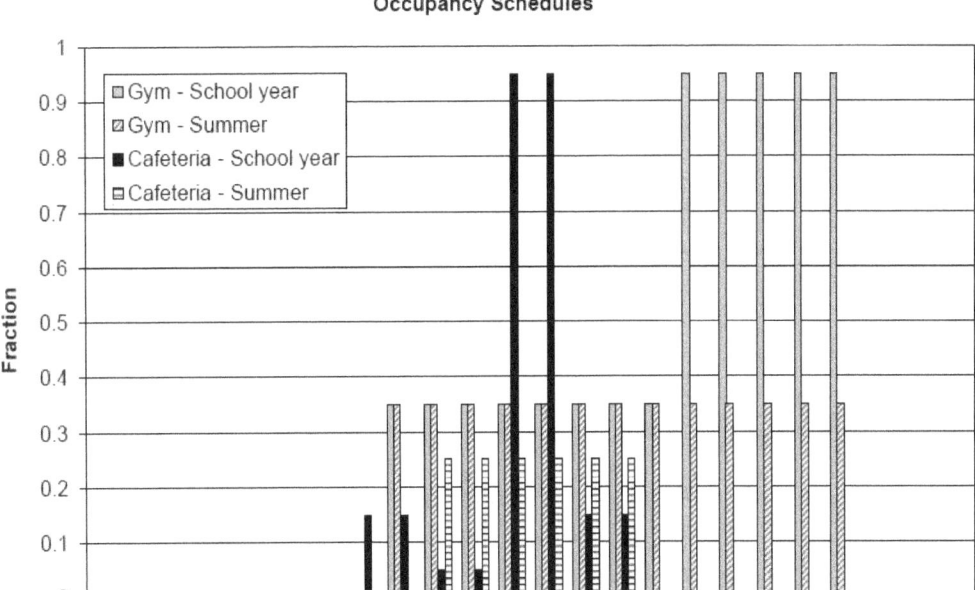

Figure A12 Occupancy schedules for Primary School (Gym, Cafeteria)

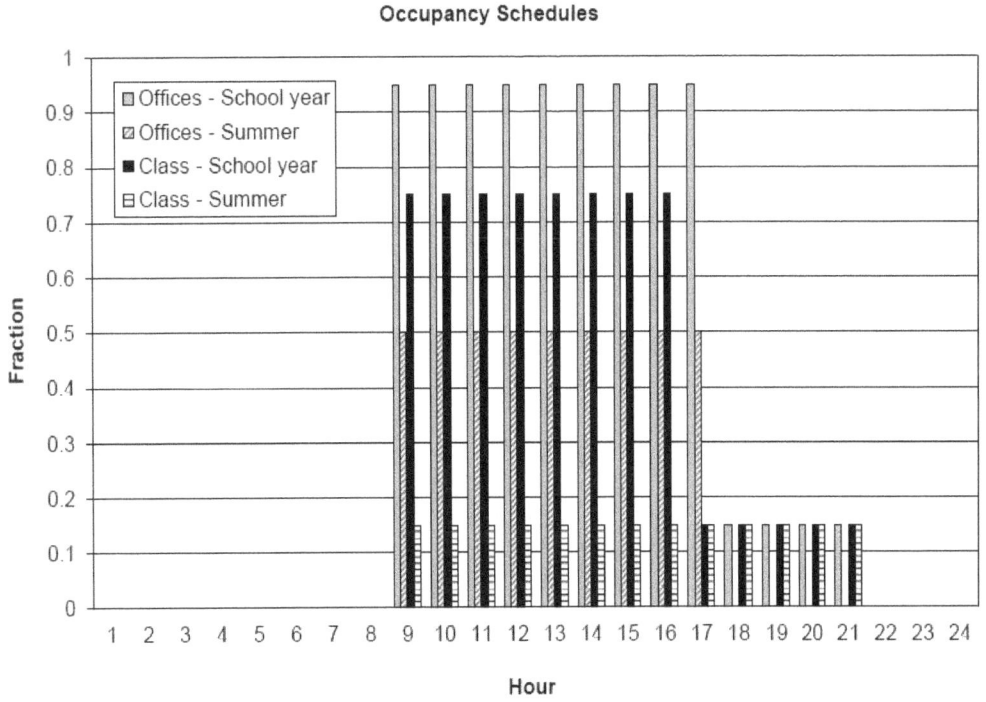

Figure A13 Occupancy schedules for Primary School (Offices, Class)

A2.7.   Secondary School
Table A14 summarizes the zones modeled in CONTAM for the Secondary School, their respective sizes, and maximum occupancy.

Geometry:
19 592 m^2 footprint ("E" shape), two-story building with flat roof. The EnergyPlus model has 25 zones on the first floor and 21 zones on the second floor. The first floor has an 11 902 m^2 footprint. The second floor is stacked on top of the first floor with a footprint of 7690 m^2. Both floors have the same floor plan except that the Library Media Center on the second floor stacks on top of the Cafeteria and Kitchen. Most of the large spaces that are typical of schools are modeled as individual zones. Some of the classrooms are grouped together to form a single thermal zone. Three spaces (Auditorium, Gym, and Auxiliary Gym) on the first floor are two-stories (8 m) high. The remaining zones on both floors are all 4 m high.

The CONTAM model was altered to create more realistic corridors with reasonable circulation patterns. In EnergyPlus, the first and second floor Bathroom blocked access to the Pod 3 Corridor. This also prevented access to those sections of the building from the first and second floor Main Corridor. In CONTAM, a 3 m wide path was carved out of the first and second floor Bathroom to provide access from the Main Corridor to the Pod 3 (shaded in Figure A14 and Figure A15).

In EnergyPlus, the first floor Mechanical Room blocked access from the Main Corridor to the Cafeteria, Kitchen, and Auxiliary Gym. In EnergyPlus, the second floor Mechanical Room blocked access from the Main Corridor to the Library Media Center. In CONTAM, the first and second floor Mechanical Room were moved and carved out of the Gym (shaded in Figure A14 and Figure A15). In CONTAM, the first floor Kitchen was also shortened to provide access from the Main Corridor to the Cafeteria and Auxiliary Gym (shaded in Figure A14).

These changes make the first and second floor Main Corridor larger (shaded in Figure A14 and Figure A15) and the first and second floor Bathroom, first floor Kitchen, and first floor Gym smaller in the CONTAM model than in the EnergyPlus model. Nevertheless, the occupancies and ventilation rates of the zones were not changed so that the CONTAM model matches the EnergyPlus model.

Table A14 Summary of zones in Secondary School

Zone	Floor	Area (m^2)	Height (m)	Maximum occupancy
Corner ("Cor") Class 1 Pod 1	1	99	4.0	24.75
Multiple ("Mult") Class 1 Pod 1	1	477	4.0	119.25
Corridor ("Corr") Pod 1	1	320	4.0	32.00
Cor Class 2 Pod 1	1	99	4.0	24.75
Mult Class 2 Pod 1	1	477	4.0	119.25
Cor Class 1 Pod 2	1	99	4.0	24.75
Mult Class 1 Pod 2	1	477	4.0	119.25
Corr Pod 2	1	320	4.0	32.00
Cor Class 2 Pod 2	1	99	4.0	24.75
Mult Class 2 Pod 2	1	477	4.0	119.25
Cor Class 1 Pod 3	1	99	4.0	24.75
Mult Class 1 Pod 3	1	477	4.0	119.25
Corr Pod 3	1	320	4.0	32.00
Cor Class 2 Pod 3	1	99	4.0	24.75
Mult Class 2 Pod 3	1	477	4.0	119.25
Main Corridor	1	1524	4.0	0.00
Lobby	1	210	4.0	0.00
Bathroom	1	195	4.0	21.00
Offices	1	532	4.0	26.60
Gym	1	1976	8.0	1976.00
Auxiliary Gym	1	1248	8.0	374.8
Auditorium	1	988	8.0	988.00
Kitchen	1	189	4.0	32.4
Cafeteria	1	624	4.0	448.92
Mechanical Room	1	342	4.0	3.42
Cor Class 1 Pod 1	2	99	4.0	24.75
Mult Class 1 Pod 1	2	477	4.0	119.25
Corr Pod 1	2	320	4.0	32.00
Cor Class 2 Pod 1	2	99	4.0	24.75
Mult Class 2 Pod 1	2	477	4.0	119.25
Cor Class 1 Pod 2	2	99	4.0	24.75
Mult Class 1 Pod 2	2	477	4.0	119.25
Corr Pod 2	2	320	4.0	32.00
Cor Class 2 Pod 2	2	99	4.0	24.75
Mult Class 2 Pod 2	2	477	4.0	119.25
Cor Class 1 Pod 3	2	99	4.0	24.75
Mult Class 1 Pod 3	2	477	4.0	119.25
Corr Pod 3	2	320	4.0	32.00
Cor Class 2 Pod 3	2	99	4.0	24.75
Mult Class 2 Pod 3	2	477	4.0	119.25
Main Corridor	2	1497	4.0	0.00
Lobby	2	210	4.0	0.00
Bathroom	2	195	4.0	21.00
Offices	2	532	4.0	26.60
Library Media Center	2	840	4.0	193.10
Mech	2	342	4.0	3.42

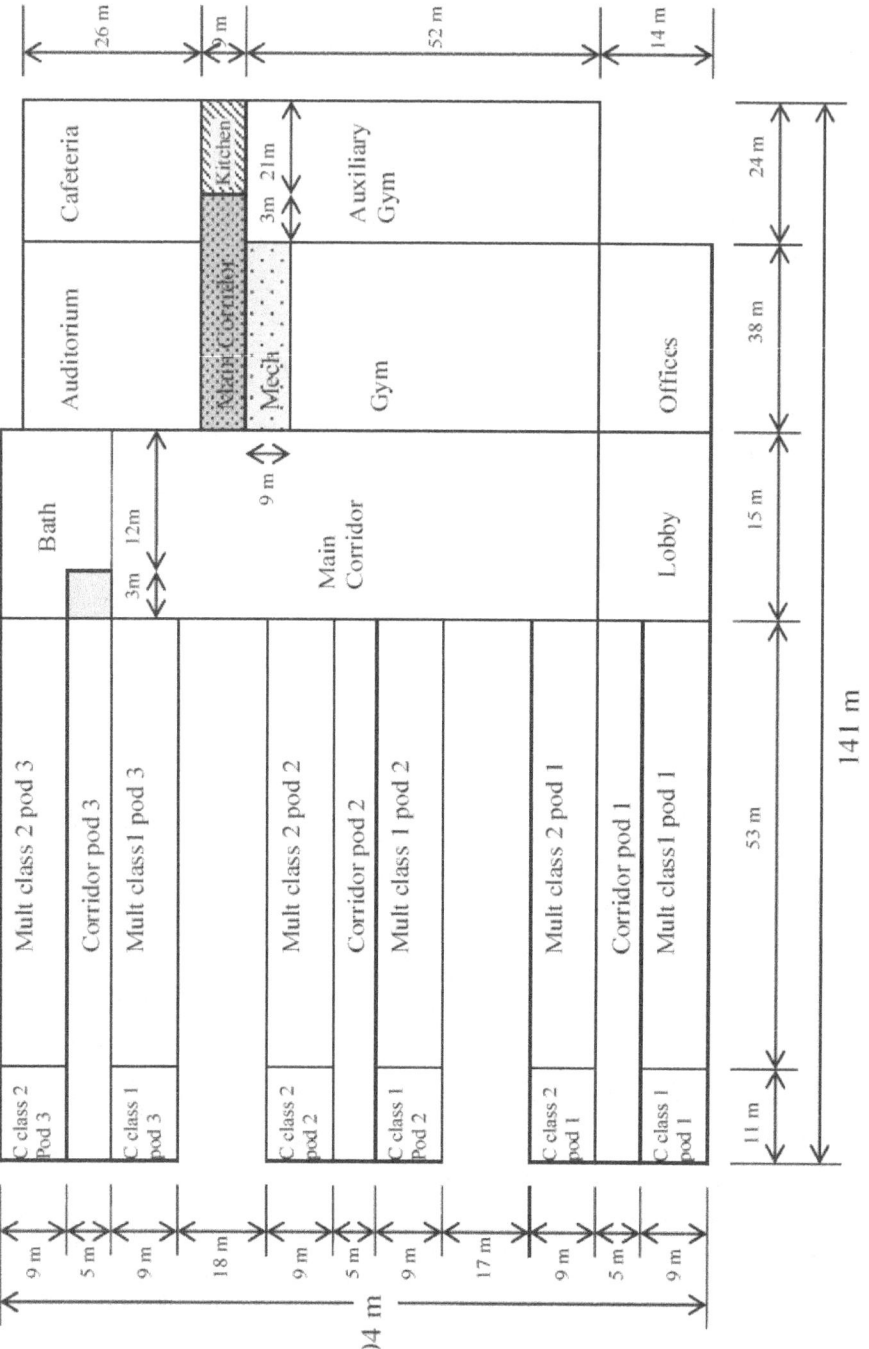

Figure A14 First floor plan of Secondary School (height 4.0 m)

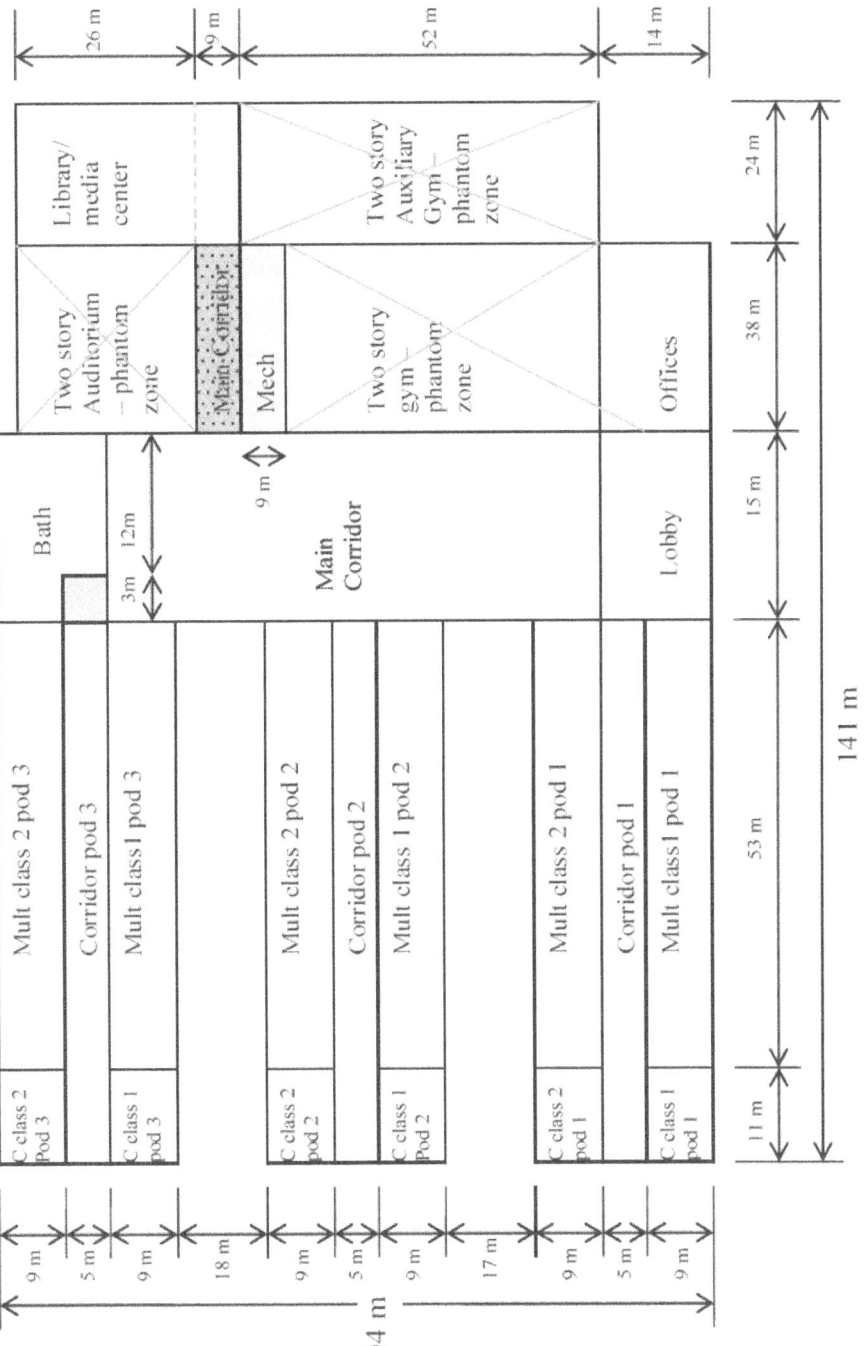

Figure A15 Second floor plan of Secondary School (height 4.0 m)

87

Large interior leakage paths were defined as follows:
- Between the Kitchen and Cafeteria zones, a single large leakage path of 42 m^2 (50 % of the total wall area between the two spaces) is modeled;
- The wall between each Pod's corridor and the Main Corridor is modeled with a 4.0 m^2 open doorway;
- Two 4.0 m^2 open doorways are modeled in the wall connecting the Lobby to the Main Corridor;
- Each Bathroom is modeled with a 0.186 m^2 transfer grille in the door between the Bathroom and the Main Corridor.

HVAC systems:
For all building vintages, the EnergyPlus model has five CAV and four VAV systems. The Gym, Auxiliary Gym, Auditorium, Kitchen, and Cafeteria each have a CAV system. The VAV systems are zoned as follows:
- VAV Pod 1: serves the zones in Pod 1, first and second floor
- VAV Pod 2: serves the zones in Pod 2, first and second floor
- VAV Pod 3: serves the zones in Pod 3, first and second floor
- VAV Other: serves the first and second floors Main Corridor, Lobby, Mechanical Room, Bathroom, Offices, and the second floor Library Media Center (11 zones total)

The design supply flow rate calculated by EnergyPlus for each VAV system is used as the supply flow rate for each *constant*-volume system modeled in CONTAM for simplicity. The systems modeled in CONTAM are still referred to as "VAV" systems in the body of this text. Varying the supply flow rate can be implemented in CONTAM using controls and/or schedules by users who wish to do so. The supply air, return air, outside air, and exhaust flow rates modeled in CONTAM are listed in Table A15 for all three building vintages.

In EnergyPlus, there was a Cafeteria exhaust fan (1.63 m^3/s) in addition to the Kitchen exhaust fan (0.26 m^3/s). It was included in order to transfer air from the Cafeteria to the Kitchen. This is modeled in CONTAM using a large opening between the Cafeteria and Kitchen zones (see above), and one larger exhaust fan in the Kitchen (1.89 m^3/s).

In EnergyPlus, neutral building pressurization is modeled in all zones. To pressurize the building in CONTAM, less air is returned than is supplied to each zone. For all building vintages, the return airflow rate is set to 90 % of the supply airflow rate. The return airflows from the first and second floor Bathrooms are reduced by the Bathroom exhausts (0.3 m^3/s each). The return air from the Kitchen is equal to the supply airflow rate minus the outside air requirement, and the return air for the Cafeteria is reduced by 1.63 m^3/s, to allow makeup air for the Kitchen exhaust (1.89 m^3/s). The Kitchen is thus neutrally pressurized.

Table A15 Summary of HVAC system flow rates (m³/s) in Secondary School

Zone	Floor	New		Post-1980		Pre-1980		Outside Air (m³/s)	Exhaust air (m³/s)
		Supply	Return	Supply	Return	Supply	Return		
Corner ("Cor") Class 1 Pod 1	1	0.44	0.39	0.49	0.44	0.57	0.51		0
Multiple ("Mult") Class 1 Pod 1	1	0.95	0.86	0.99	0.89	1.18	1.06		0
Corridor ("Corr") Pod 1	1	0.41	0.37	0.42	0.37	0.46	0.42		0
Cor Class 2 Pod 1	1	0.41	0.36	0.46	0.41	0.53	0.48		0
Mult Class 2 Pod 1	1	0.95	0.86	0.95	0.86	1.02	0.92		0
Cor Class 1 Pod 1	2	0.64	0.57	0.67	0.60	0.81	0.73		0
Mult Class 1 Pod 1	2	1.64	1.48	1.72	1.55	2.14	1.93		0
Corr Pod 1	2	0.78	0.70	0.77	0.69	0.98	0.88		0
Cor Class 2 Pod 1	2	0.61	0.55	0.64	0.58	0.78	0.70		0
Mult Class 2 Pod 1	2	1.48	1.33	1.57	1.41	1.88	1.69		0
VAV-Pod 1 Total		**8.31**	**7.48**	**8.68**	**7.81**	**10.35**	**9.31**	**4.93**	
Cor Class 1 Pod 2	1	0.43	0.38	0.48	0.43	0.56	0.51		0
Mult Class 1 Pod 2	1	0.95	0.86	1.09	0.98	1.27	1.14		0
Corr Pod 2	1	0.29	0.26	0.35	0.32	0.40	0.36		0
Cor Class 2 Pod 2	1	0.40	0.36	0.46	0.41	0.53	0.48		0
Mult Class 2 Pod 2	1	0.95	0.86	0.95	0.86	1.02	0.92		0
Cor Class 1 Pod 2	2	0.63	0.57	0.66	0.60	0.80	0.72		0
Mult Class 1 Pod 2	2	1.61	1.45	1.71	1.54	2.12	1.91		0
Corr Pod 2	2	0.77	0.69	0.76	0.69	0.97	0.88		0
Cor Class 2 Pod 2	2	0.61	0.55	0.64	0.58	0.78	0.70		0
Mult Class 2 Pod 2	2	1.48	1.33	1.58	1.42	1.89	1.70		0
VAV-Pod 2 Total		**8.13**	**7.32**	**8.70**	**7.83**	**10.35**	**9.32**	**4.93**	
Cor Class 1 Pod 3	1	0.43	0.38	0.48	0.43	0.56	0.51		0
Mult Class 1 Pod 3	1	0.95	0.86	1.09	0.98	1.28	1.15		0
Corr Pod 3	1	0.29	0.26	0.36	0.32	0.41	0.36		0
Cor Class 2 Pod 3	1	0.41	0.37	0.46	0.42	0.54	0.49		0
Mult Class 2 Pod 3	1	0.95	0.86	0.98	0.88	1.06	0.95		0
Cor Class 1 Pod 3	2	0.64	0.57	0.67	0.60	0.81	0.73		0
Mult Class 1 Pod 3	2	1.91	1.72	2.00	1.80	2.42	2.18		0
Corr Pod 3	2	0.78	0.70	0.78	0.70	0.99	0.89		0
Cor Class 2 Pod 3	2	0.62	0.56	0.65	0.58	0.78	0.70		0
Mult Class 2 Pod 3	2	1.79	1.61	1.88	1.69	2.20	1.98		0

Zone	Floor	New		Post-1980		Pre-1980		Outside	Exhaust
VAV-Pod 3 Total		**8.77**	**7.89**	**9.34**	**8.41**	**11.04**	**9.94**	**4.93**	
Main Corridor	1	1.14	1.03	1.80	1.62	2.00	1.80		0
Lobby	1	0.24	0.22	0.32	0.28	0.35	0.31		0
Bathroom	1	0.53	0.17	0.70	0.33	0.76	0.38		0.30
Offices	1	0.91	0.82	1.14	1.03	1.30	1.17		0
Mechanical Room	1	1.05	0.95	0.89	0.80	0.91	0.82		0
Main Corridor	2	2.81	2.53	2.77	2.49	3.58	3.22		0
Lobby	2	0.54	0.48	0.60	0.54	0.68	0.61		0
Bathroom	2	0.65	0.28	0.86	0.48	0.98	0.58		0.30
Offices	2	1.66	1.49	1.77	1.59	2.19	1.97		0
Library Media Center	2	2.53	2.28	2.45	2.21	2.97	2.67		0
Mechanical Room	2	0.76	0.68	0.51	0.46	0.68	0.61		0
VAV-Other Total		**12.80**	**10.92**	**13.81**	**11.83**	**16.39**	**14.15**	**4.65**	
Gym (CAV 1:5)	1	19.79	17.78	19.79	17.78	19.79	17.78	19.76	0
Auxiliary Gym (CAV 2:6)	1	4.97	4.47	3.75	3.38	4.43	3.99	3.75	0
Auditorium (CAV 3:7)	1	7.90	7.11	7.90	7.11	7.90	7.11	7.90	0
Kitchen (CAV 4:8)	1	0.55	0.29	0.56	0.30	0.59	0.33	0.26	1.89
Cafeteria (CAV 5:9)	1	4.49	2.41	4.49	2.41	4.49	2.41	4.49	0

Schedules:

All HVAC and exhaust fans operate on the following schedule:

- Weekdays: on from 6:00-21:00, off otherwise
- Weekends and holidays: off all day

Outside air is supplied according to this schedule as well. Summer weekday schedules, as discussed in Section A1, are not used in the HVAC/fan schedules.

Occupants:

The peak number of people for each zone is listed in Table A14. Occupants in all building zones are scheduled according to Figure A16 to Figure A17. Sundays and holidays are unoccupied. There are six different occupant schedules for the building. The occupancy schedules for the Gym (and Auxiliary Gym), Cafeteria, and Auditorium are shown in Figure A16. The occupancy schedule for the Offices is shown in Figure A17. Also shown in Figure A17 is the occupancy schedule for the remaining zones (referred to as "Class" occupancy). There is also an *Extended* Class occupancy schedule in Figure A17. They are for the "Mult Class 1" "Mult Class 2" zones in Pod 3 on the second floor. The Extended Class and Class occupancy schedules are the same except the Extended Class occupancy schedule has greater and constant occupancy from 9:00 a.m. to 9:00 p.m. both during the school year and summer. All of the schedules in Figure A16 to Figure A17 consider "summer" to be July 1 through August 31. "School year" is the remainder of the year.

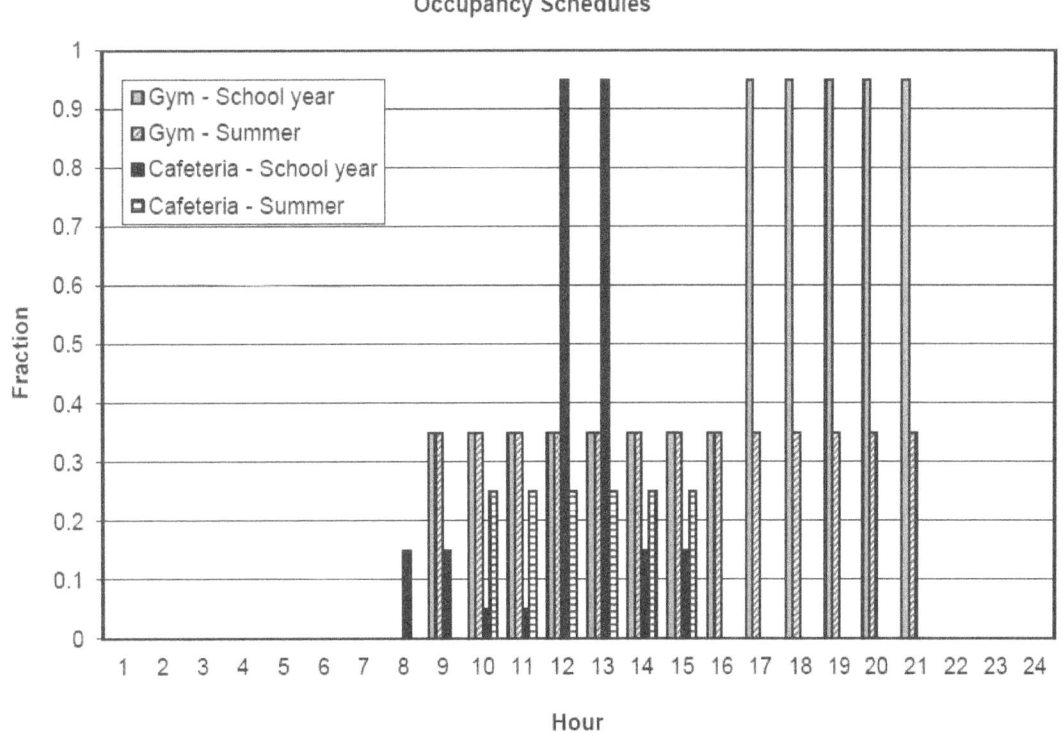

Figure A16 Occupancy schedules for Secondary School (Gym, Cafeteria, and Auditorium)

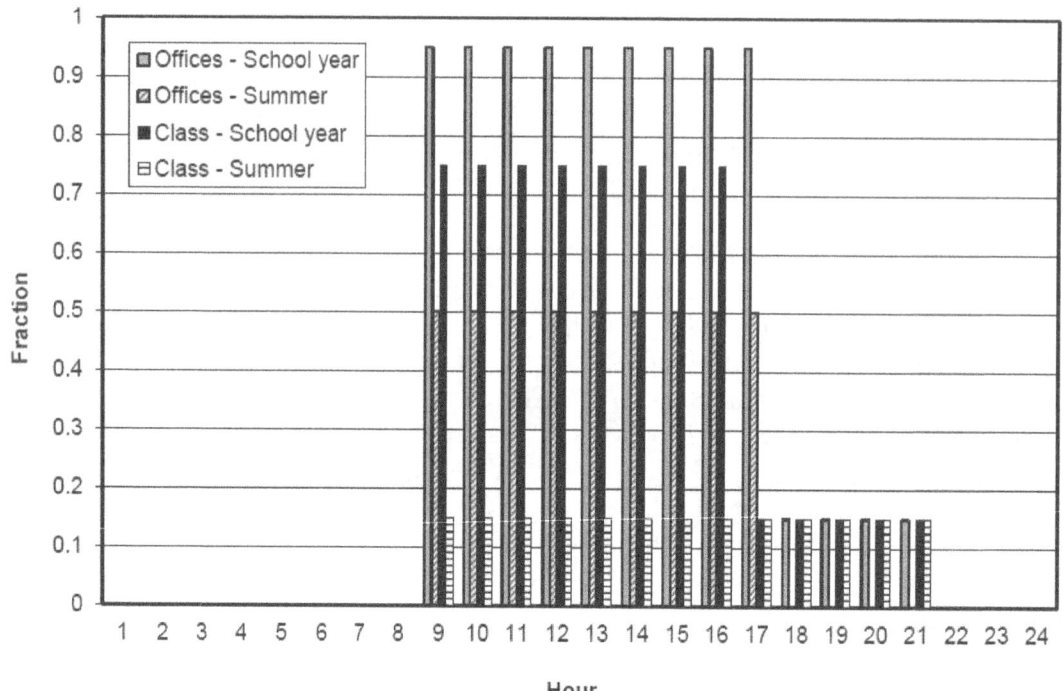

Figure A17 Occupancy schedules for Secondary School (Offices, Class)

A2.8. Stand-Alone Retail

Table A16 summarizes the zones modeled in CONTAM for the Stand-Alone Retail, their respective sizes, and maximum occupancy.

Table A16 Summary of zones in Stand-Alone Retail

Zone	Area (m^2)	Height (m)	Maximum occupancy
Back Space	352	6.1	13.63
Core Retail	1600	6.1	258.40
Point of Sale	151	6.1	24.35
Front Retail	151	6.1	24.35
Front Entry	12	6.1	1.94
Restroom	28	6.1	0

Geometry:

2294 m^2 footprint, single-story building with flat roof. The EnergyPlus model has five zones. In the CONTAM model, a Restroom (shaded in Figure A18) with a footprint of 4 m × 7 m was carved out of the Back Space.

92

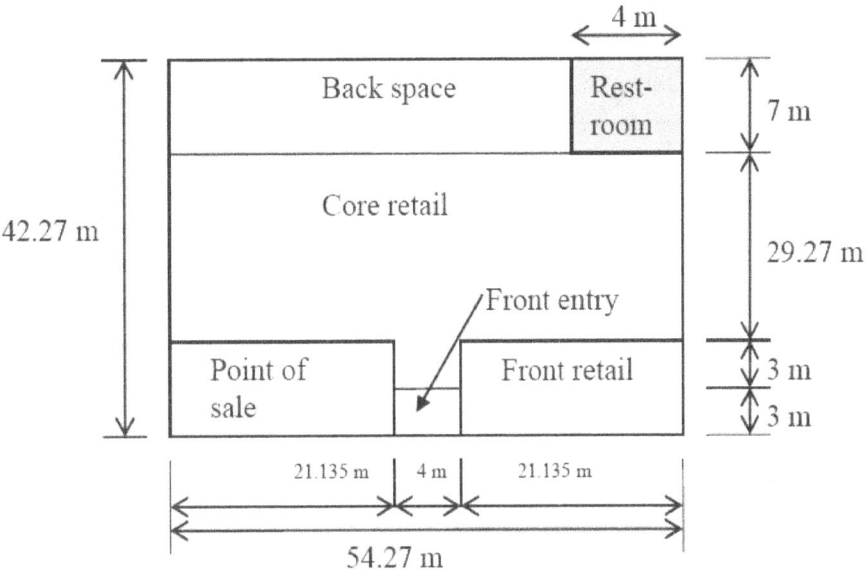

Figure A18 Floor plan of Stand-Alone Retail (height 6.1 m)

Large interior leakage paths were defined as follows:
- Between Back Space and Core Retail, a single large leakage path of 3.0 m² is modeled;
- Between Core Retail and Point of Sale, a single large leakage path of 96.7 m² (75 % of the total wall area between the spaces) is modeled;
- Between Core Retail and Front Retail, a single large leakage path of 96.7 m² (75 % of the total wall area between the spaces) is modeled;
- Between the Restroom and Back space, a 0.372 m² transfer grille is modeled.

HVAC systems:
For all building vintages, the building has five packaged constant-volume single-zone systems. Similarly, each zone has a constant-volume system in CONTAM. The supply air, return air, outside air, and exhaust flow rates modeled in CONTAM are listed in Table A17 for all three building vintages. The exhaust flow rate for the Restroom was modeled only in CONTAM, not in EnergyPlus.

The Front Entry has a unit heater in EnergyPlus that recirculates air locally within the zone and does not impact whole-building airflow or introduce outside air. Therefore, the unit heater is not modeled in CONTAM.

In EnergyPlus, neutral building pressurization is modeled in all zones. To pressurize the building in CONTAM, less air is returned than is supplied to each zone. For all building vintages, the return airflow rate is set to 90 % of the supply airflow rate.

Table A17 Summary of HVAC system flow rates (m³/s) in Stand-Alone Retail

Zone	New		Post-1980		Pre-1980		Outside Air (m³/s)	Exhaust air (m³/s)
	Supply	Return	Supply	Return	Supply	Return		
Back Space	1.44	1.30	3.31	3.03	3.57	3.29	0.29	0
Core Retail	4.72	4.25	7.21	6.49	7.69	6.92	2.40	0
Point of Sale	0.88	0.79	1.47	1.32	1.58	1.42	0.23	0
Front Retail	0.66	0.59	1.47	1.32	1.58	1.42	0.23	0
Restroom	N/A	N/A	N/A	N/A	N/A	N/A	0	0.05

Schedules:

All HVAC and exhaust fans operate on the following schedule:

- Weekdays: on from 6:00 a.m. to 9:00 p.m., off otherwise
- Saturday: on from 6:00 a.m. to 10:00 p.m., off otherwise
- Sunday, holidays: on from 8:00 a.m. to 7:00 p.m., off otherwise

Outside air is supplied according to this schedule as well.

Occupants:

The peak number of people for each zone is listed in Table A16. Occupants in all building zones are scheduled according to Figure A19. There is a different occupancy schedule for weekdays, Saturdays, and Sundays/ holidays.

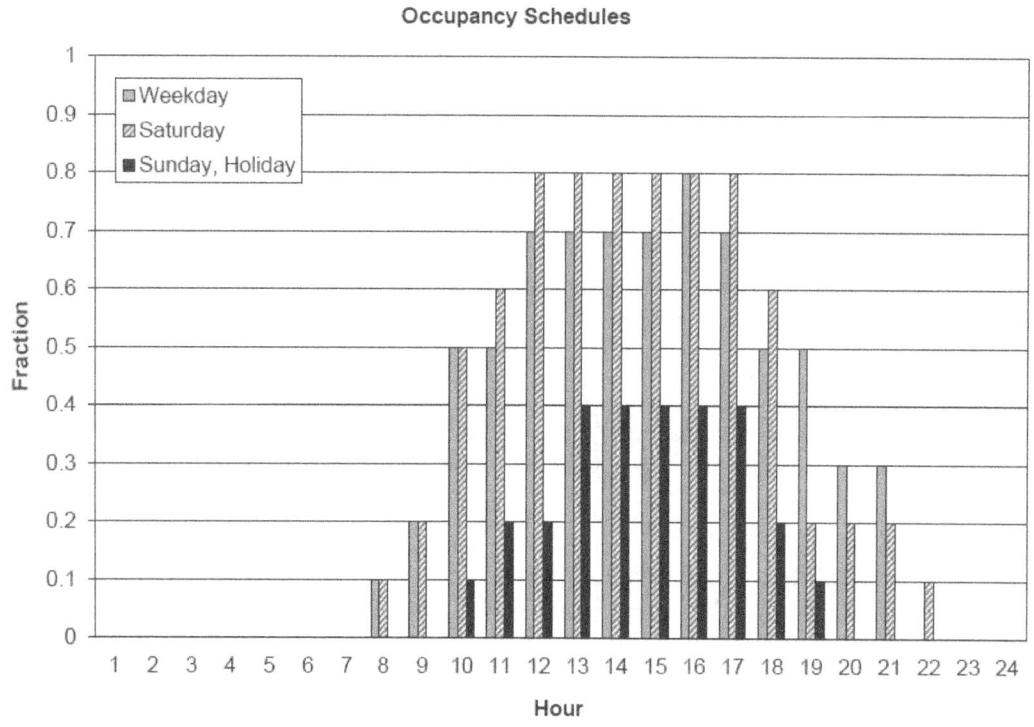

Figure A19 Occupancy schedule for Stand-Alone Retail

A2.9. Strip Mall

Table A18 summarizes the zones modeled in CONTAM for the Strip Mall, their respective sizes, and maximum occupancy.

Table A18 Summary of zones in Strip Mall

Zone	Area (m²)	Height (m)	Maximum occupancy
Large Stores (LS)	261.29	5.18	56.25
LS storage	52.81	5.18	0
LS restroom	34.29	5.18	0
Small Stores (SS)	130.64	5.18	28.12
SS storage	39.55	5.18	0
SS restroom	4.00	5.18	0

Geometry:

2090 m² footprint, single-story building with flat roof. The EnergyPlus model has 10 zones. There are two "Large Stores" (LS) and eight "Small Stores" (SS). In the CONTAM model, each store is sub-divided into three zones. The rear 25 % of each store was divided into Storage and Restroom zones (Storage zones are shown in gray and Restrooms shown in black in Figure A20).

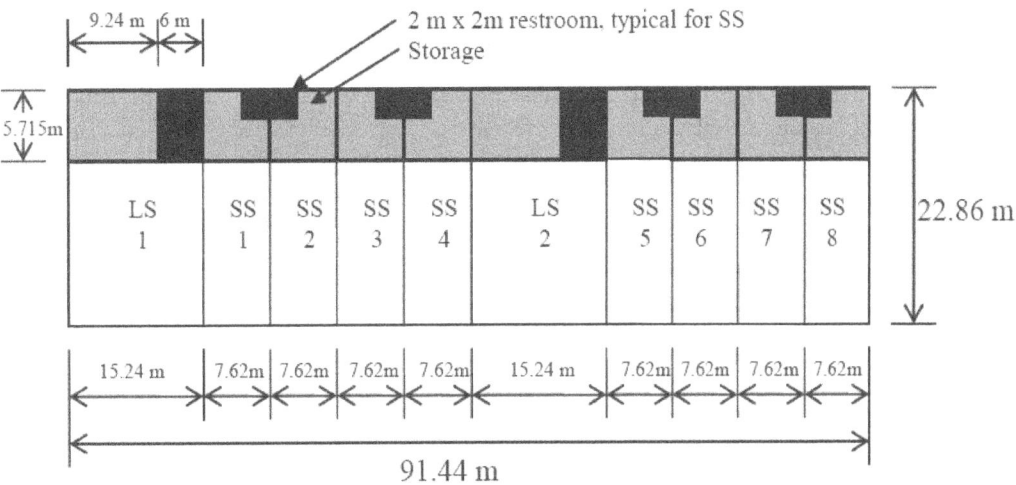

Figure A20 Floor plan of Strip Mall (height 5.18 m)

Large interior leakage paths were defined as follows:
- Between the store and storage zones, a 4 m² opening is modeled;
- Between the Restrooms in the Large Stores and Storage zones, a 0.186 m² transfer grille is modeled;
- Between the Restrooms in the Small Stores and Storage zones, a 0.025 m² door undercut is modeled.

HVAC systems:

For all building vintages, the EnergyPlus model has 10 packaged constant-volume single-zone systems. Similarly, each zone has a constant-volume system in CONTAM. The supply air, return air, outside air, and exhaust flow rates modeled in CONTAM are listed in Table A19 for all three building vintages. The exhaust flow rates for the Restrooms were modeled only in CONTAM, not in EnergyPlus.

In CONTAM, the design supply flow rate calculated by EnergyPlus for each constant-volume system is split between the store and storage zones. Seventy five percent of the design supply flow rate is directed to the store and 25 % to the storage zone in CONTAM. In EnergyPlus, neutral building pressurization is modeled in all zones. To pressurize the building in CONTAM, less air is returned than is supplied to each zone. For all building vintages, the return plus Restroom exhaust flow rate for each zone is set to 90 % of the supply airflow rate.

Table A19 Summary of HVAC system flow rates (m³/s) in Strip Mall

Zone	New		Post-1980		Pre-1980		Outside Air (m³/s)	Exhaust air (m³/s)
	Supply	Return	Supply	Return	Supply	Return		
LS 1 (CAV 1:1)	1.38	1.22	2.50	2.23	2.72	2.42	0.52	0
LS 2 (CAV 6:6)	1.02	0.89	2.01	1.78	2.16	1.92	0.52	0
SS 1 (CAV 2:2)	0.65	0.56	1.19	1.05	1.25	1.10	0.26	0
SS 2 (CAV 3:3)	0.59	0.51	1.01	0.88	1.09	0.96	0.26	0
SS 3 (CAV 4:4)	0.59	0.51	1.01	0.88	1.09	0.96	0.26	0
SS 4 (CAV 5:5)	0.58	0.50	1.01	0.88	1.09	0.96	0.26	0
SS 5 (CAV 7:7)	0.53	0.45	1.01	0.88	1.09	0.96	0.26	0
SS 6 (CAV 8:8)	0.53	0.45	1.01	0.88	1.09	0.96	0.26	0
SS 7 (CAV 9:9)	0.53	0.45	1.01	0.88	1.09	0.96	0.26	0
SS 8 (CAV 10:10)	0.60	0.52	1.50	1.33	1.65	1.46	0.26	0
LS Restroom	N/A	N/A	N/A	N/A	N/A	N/A	0	0.03
SS Restroom	N/A	N/A	N/A	N/A	N/A	N/A	0	0.03

Note: The supply and return rates are the sum of the rates to/from the store and storage zones.

Schedules:

All HVAC and exhaust fans operate on the following schedule:
- Weekdays: on from 6:00 a.m. to 9:00 p.m., off otherwise
- Saturday: on from 6:00 a.m. to 10:00 p.m., off otherwise
- Sunday, holidays: on from 8:00 a.m. to 7:00 p.m., off otherwise

Outside air is supplied according to this schedule as well.

Occupants:

The peak number of people for each zone is listed in Table A18. Occupants in all building zones are scheduled according to Figure A21. There is a different occupancy schedule for weekdays, Saturdays, and Sundays/ holidays.

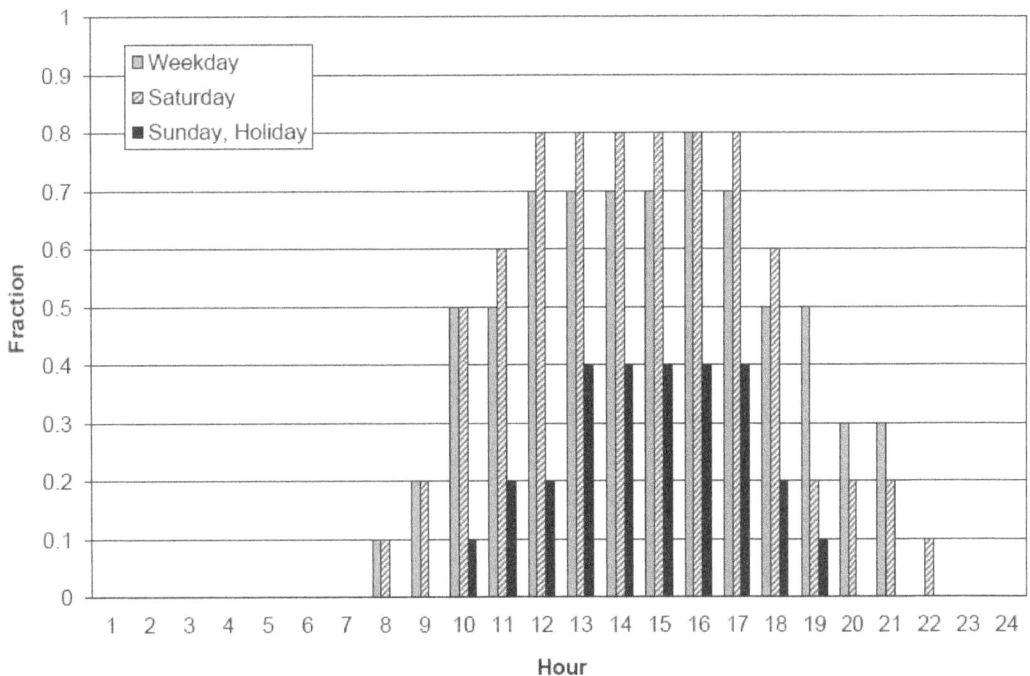

Figure A21 Occupancy schedule for Strip Mall

A2.10. Supermarket

Table A20 summarizes the zones modeled in CONTAM for the Supermarket, their respective sizes, and maximum occupancy.

Geometry:

4181 m^2 footprint, single-story building with flat roof. The EnergyPlus model has six zones. In the CONTAM model, a Restroom (shaded in Figure A22) with a footprint of 4 m × 4 m was carved out of Dry Storage.

Table A20 Summary of zones in Supermarket

Zone	Area (m^2)	Height (m)	Maximum occupancy
Office	89	6.10	4.78
Dry Storage	606	6.10	22.31
Deli	225	6.10	19.35
Sales	2325	6.10	200.20
Produce	711	6.10	61.26
Bakery	209	6.10	18.00
Restroom	16	6.10	0

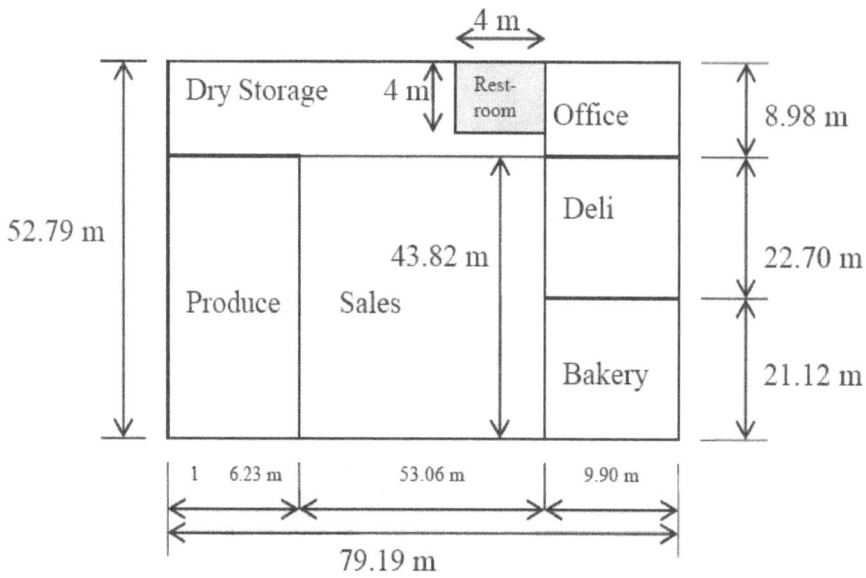

Figure A22 Floor plan of Supermarket (height 6.1 m)

Large interior leakage paths were defined as follows:
- Between Produce and Sales, a single large leakage path of 6.0 m^2 is modeled;
- Between Bakery and Sales, a single large leakage path of 64.4 m^2 (50 % of the total wall area between the spaces) is modeled;
- Between Deli and Sales, a single large leakage path of 69.2 m^2 (50 % of the total wall area between the spaces) is modeled;
- Between Dry Storage and Sales zone, a single large leakage path of 6.0 m^2 is modeled;
- Between Restroom and Dry Storage zone, a 0.186 m^2 transfer grille is modeled.

HVAC systems:
For all building vintages, the EnergyPlus model has five packaged constant-volume single-zone systems. Similarly, each zone has a constant-volume system in CONTAM. The supply air, return air, outside air, and exhaust flow rates modeled in CONTAM are listed in Table A21 for all three building vintages. The exhaust flow rates for the Restrooms were modeled only in CONTAM, not in EnergyPlus.

In EnergyPlus, there was a Sales exhaust fan (1.08 m^3/s) in addition to the Bakery exhaust fan (0.35 m^3/s) and Deli exhaust fan (0.34 m^3/s). The Sales exhaust fan was included in order to transfer air from the Sales zone to the Deli. This is modeled in CONTAM using a large opening between the Sales and Deli zones (see above), and one larger exhaust fan in the Deli (1.42 m^3/s). The sum of the Sales and Deli exhaust rates in EnergyPlus is 1.42 m^3/s.

In EnergyPlus, neutral building pressurization is modeled in all zones. To pressurize the building in CONTAM, less air is returned than is supplied to each zone. For the Dry Storage, Office, and Produce zones, the return airflow rate is the larger of (a) 90 % of the supply airflow rate and (b) supply airflow rate minus outside air requirement.

In CONTAM, the Bakery is neutrally pressurized. The return airflow rate is the supply airflow rate minus the exhaust rate for all building vintages.

In CONTAM, the Deli is depressurized. The return airflow rate is the supply airflow rate minus the outside air rate for all building vintages. The Deli receives 1.08 m^3/s of air from Sales, as modeled in EnergyPlus (see above). Thus, for Sales, the return airflow rate is the supply airflow rate minus 1.08 m^3/s for all building vintages.

Table A21 Summary of HVAC system flow rates (m^3/s) in Supermarket

Zone	New		Post-1980		Pre-1980		Outside Air (m^3/s)	Exhaust air (m^3/s)
	Supply	Return	Supply	Return	Supply	Return		
Office	0.58	0.53	1.05	1.00	1.13	1.08	0.05	0
Dry Storage	3.06	2.73	5.63	5.16	6.03	5.56	0.47	0
Deli	2.17	1.83	2.39	2.05	2.33	1.99	0.34	1.42
Sales	9.69	7.60	16.05	13.33	16.56	13.78	3.49	0
Produce	3.20	2.88	5.74	5.17	6.07	5.46	1.07	0
Bakery	2.14	1.83	2.33	2.02	2.28	1.97	0.31	0.35
Restroom	N/A	N/A	N/A	N/A	N/A	N/A	0	0.03

Schedules:

All HVAC and exhaust fans operate on the following schedule:
- Everyday: on from 6:00 a.m. to 10:00 p.m., off otherwise

Outside air is supplied according to this schedule as well.

Occupants:

The peak number of people for each zone is listed in Table A20. Occupants in all building zones are scheduled according to Figure A23. There is a different occupancy schedule for weekdays, Saturdays, and Sundays/ holidays.

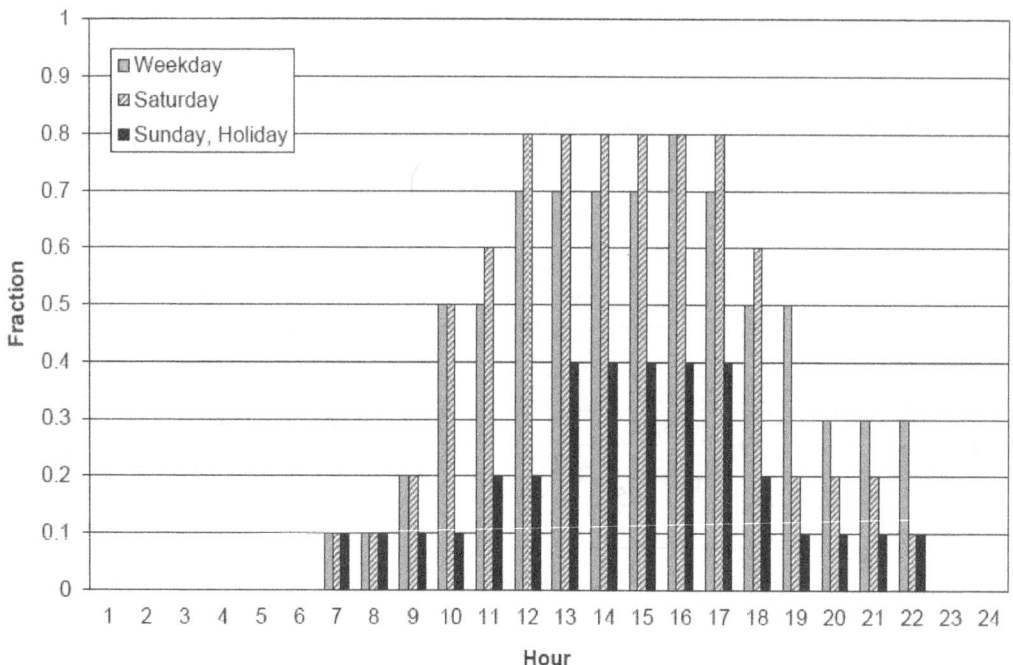

Figure A23 Occupancy schedule for Supermarket

A2.11.   Small Hotel

Table A22 summarizes the zones modeled in CONTAM for the Small Hotel, their respective sizes, and maximum occupancy.

Geometry:

1003 m^2 footprint, four-story building with flat roof. Total floor area is 4014 m^2. The EnergyPlus model has 19 zones on the first floor and 16 zones on each of the upper floors (Figure A24). The upper floor plans are identical. On the upper floors, there are five zones that represent three or four hotel rooms lumped together. These are also modeled as single zones in the CONTAM model. The Post-1980 and Pre-1980 buildings also have an attic that is 1.45 m high.

Large interior leakage paths were defined as follows:
- Between Front Lounge and Corridor zones, a single large leakage path (50 % of the total wall area between the spaces) is modeled;
- Between Restroom and Corridor, a 0.186 m^2 transfer grille is modeled.

100

Table A22 Summary of zones in Small Hotel

Zone	Floor	Area (m^2)	Height (m)	Maximum occupancy
Rear Stairs	1	20	3.35	0
Corridor	1	151	3.35	0
Rear Storage	1	20	3.35	0
Front Lounge	1	163	3.35	52.71
Restroom	1	33	3.35	1.00
Meeting Room	1	80	3.35	43.20
Mechanical Room	1	33	3.35	0
Guest 101	1	33	3.35	1.41
Guest 102	1	33	3.35	1.41
Guest 103	1	33	3.35	1.41
Guest 104	1	33	3.35	1.41
Guest 105	1	33	3.35	1.41
Employee Lounge	1	33	3.35	11
Laundry	1	98	3.35	11
Elevator	1	15	3.35	0
Exercise	1	33	3.35	11
Front Office	1	130	3.35	10.03
Front Stairs	1	20	3.35	0
Front Storage	1	13	3.35	0
Rear Stairs	2-4	20	2.74	0
Corridor	2-4	125	2.74	0
Rear Storage	2-4	20	2.74	0
Guest x01	2-4	33	2.74	1.41
Guest x02-5	2-4	130	2.74	5.55
Guest x06-8	2-4	105	2.74	4.50
Guest x09-12	2-4	130	2.74	5.55
Guest x13	2-4	33	2.74	1.41
Guest x14	2-4	33	2.74	1.41
Guest x15-18	2-4	130	2.74	5.55
Elevator	2-4	15	2.74	0
Guest x19	2-4	33	2.74	1.41
Guest x20-23	2-4	130	2.74	5.55
Guest x24	2-4	33	2.74	1.41
Front Storage	2-4	13	2.74	0
Front Stairs	2-4	20	2.74	0

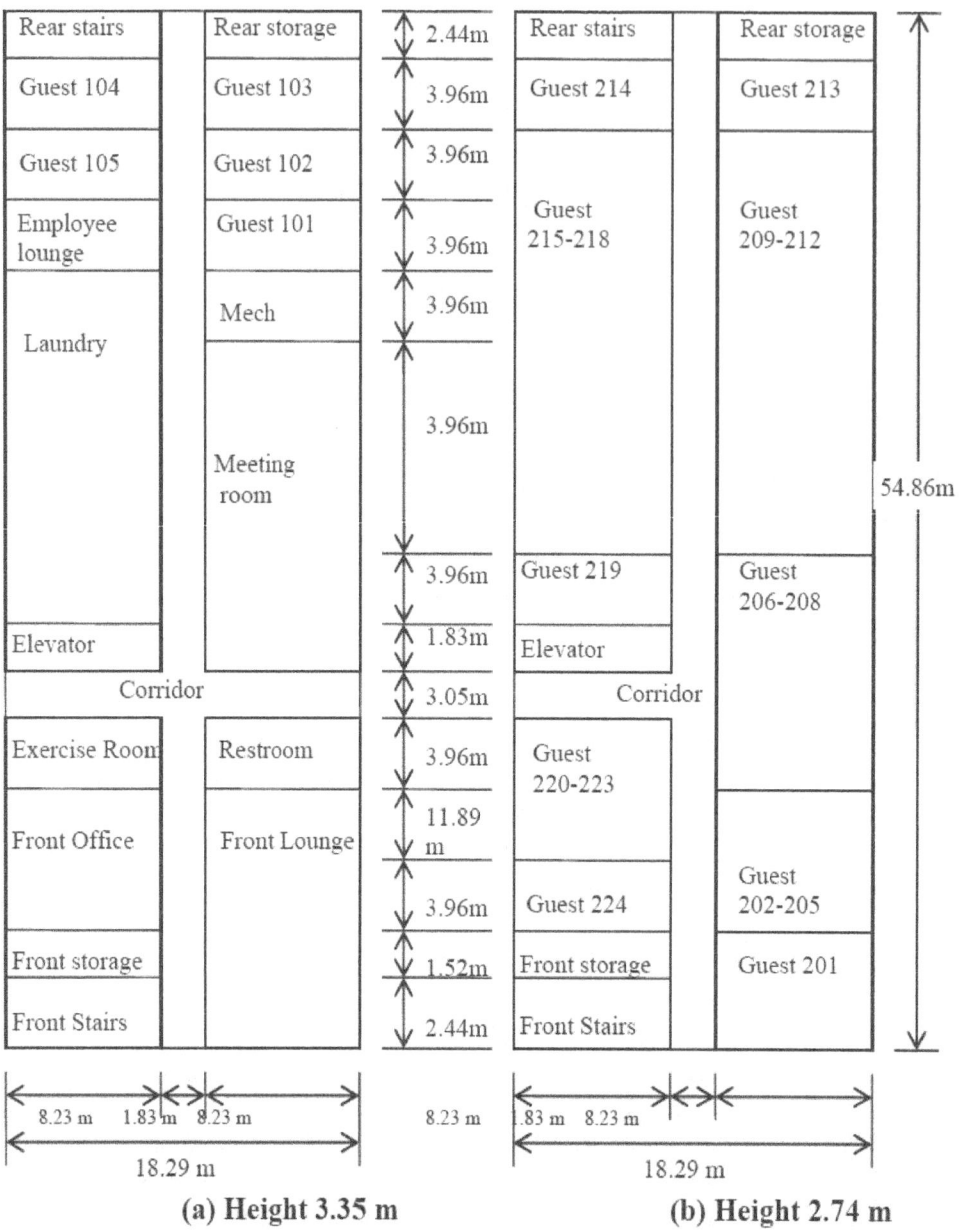

Rear stairs	Rear storage	↕ 2.44m	Rear stairs	Rear storage
Guest 104	Guest 103	↕ 3.96m	Guest 214	Guest 213
Guest 105	Guest 102	↕ 3.96m		
Employee lounge	Guest 101	↕ 3.96m	Guest 215-218	Guest 209-212
Laundry	Mech	↕ 3.96m		
	Meeting room	↕ 3.96m		
		↕ 3.96m	Guest 219	Guest 206-208
Elevator		↕ 1.83m	Elevator	
Corridor		↕ 3.05m	Corridor	
Exercise Room	Restroom	↕ 3.96m	Guest 220-223	
Front Office	Front Lounge	↕ 11.89 m		Guest 202-205
		↕ 3.96m	Guest 224	
Front storage		↕ 1.52m	Front storage	Guest 201
Front Stairs		↕ 2.44m	Front Stairs	

54.86m

| 8.23 m | 1.83 m | 8.23 m | | 8.23 m | 1.83 m | 8.23 m |

18.29 m   18.29 m

(a) Height 3.35 m   **(b) Height 2.74 m**

Figure A24 (a) First and (b) upper floor (2-4) plans of Small Hotel

HVAC systems:

For all building vintages, the EnergyPlus model has 12 packaged constant-volume single-zone systems serving the common areas. Similarly, each zone (except the first floor Restroom) has a constant-volume system in CONTAM. The constant-volume system for the Restroom in EnergyPlus does not provide ventilation. Therefore, it is not modeled in CONTAM. The Restroom exhaust depressurizes the zone. A transfer grille in the door facilitates air exchange with the adjacent Corridor (see above). Thus, the exhaust flow rate for the Restroom was modeled only in CONTAM, not in EnergyPlus.

The supply air, return air, outside air, and exhaust flow rates modeled in CONTAM are listed in Table A23 for all three building vintages. In EnergyPlus, all of the guest rooms have individual packaged-terminal air conditioning (PTAC) units that do not use an economizer. The PTAC are modeled in CONTAM as local supply fans that inject the designated amount of outside air listed in Table A24.

In EnergyPlus, the Front Stair, Rear Stair, Front Storage, and Rear Storage are heated using unit heaters that recirculate air locally within the zones and do not impact whole-building airflow or introduce outside air. Therefore, the unit heaters are not modeled in CONTAM.

In EnergyPlus, neutral building pressurization is modeled in all zones. To pressurize the building in CONTAM, less air is returned than is supplied to each zone. In CONTAM, for the common areas in Table A23, the building is pressurized by returning the larger of (a) 90 % of the supply airflow or (b) the supply airflow minus the outside air requirement. The guest rooms in Table A24 are pressurized by modeling an exhaust rate of 0.013 m^3/s for each. For example, Guest x02-05 is in reality four guest rooms, so that the exhaust rate is $0.013 \times 4 = 0.05$ m^3/s.

Table A23 Summary of HVAC system flow rates (m^3/s) in Small Hotel

Zone	Floor	New		Post-1980		Pre-1980		Outside Air (m^3/s)	Exhaust air (m^3/s)
		Supply	Return	Supply	Return	Supply	Return		
Corridor	1	0.49	0.45	0.55	0.51	0.57	0.53	0.04	0
Corridor	2	0.23	0.21	0.30	0.27	0.32	0.29	0.03	0
Corridor	3	0.21	0.19	0.28	0.25	0.30	0.27	0.03	0
Corridor	4	0.34	0.31	0.31	0.28	0.36	0.33	0.03	0
Employee Lounge	1	0.31	0.28	0.33	0.30	0.34	0.30	0.09	0
Exercise	1	0.12	0.11	0.14	0.13	0.14	0.13	0.11	0
Front Lounge	1	0.73	0.66	0.82	0.74	0.90	0.81	0.42	0
Front Office	1	0.32	0.29	0.43	0.39	0.45	0.41	0.10	0
Laundry Room	1	1.54	1.40	1.57	1.43	1.58	1.43	0.14	0
Mechanical Room	1	0.05	0.05	0.05	0.05	0.07	0.06	0.01	0
Meeting Room	1	0.43	0.39	0.47	0.43	0.50	0.45	0.43	0
Restroom	1	N/A	N/A	N/A	N/A	N/A	N/A	0	0.19

Schedules:
All HVAC and exhaust fans operate 24 hours per day every day of the year. Outside air is also supplied all of the time.

Occupants:
The peak number of people for each zone is listed in Table A22. Occupants in all building zones are scheduled according to Figure A25 to Figure A27. There are seven different occupant schedules for the building. The occupancy schedules for the Restroom and Exercise zones are shown in Figure A25. The occupancy schedule for the Lounges, Laundry, Meeting Room, and Office is shown in Figure A26. The occupancy schedule for the guestrooms (referred to as "Guest" occupancy) is shown in Figure A27. For some zones, there is a different occupancy schedule for weekdays, Saturdays, and Sundays/ holidays.

Table A24 Summary of PTAC flow rates (m^3/s) in Small Hotel

Zone	Floor	Outside Air (m^3/s)	Exhaust air (m^3/s)
Guest 101	1	0.014	0.013
Guest 102	1	0.014	0.013
Guest 103	1	0.014	0.013
Guest 104	1	0.014	0.013
Guest 105	1	0.014	0.013
Guest x01	2-4	0.014	0.013
Guest x02-5	2-4	0.06	0.05
Guest x06-8	2-4	0.042	0.038
Guest x09-12	2-4	0.06	0.05
Guest x13	2-4	0.014	0.013
Guest x14	2-4	0.014	0.013
Guest x15-18	2-4	0.06	0.05
Guest x19	2-4	0.014	0.013
Guest x20-23	2-4	0.06	0.05
Guest x24	2-4	0.014	0.013

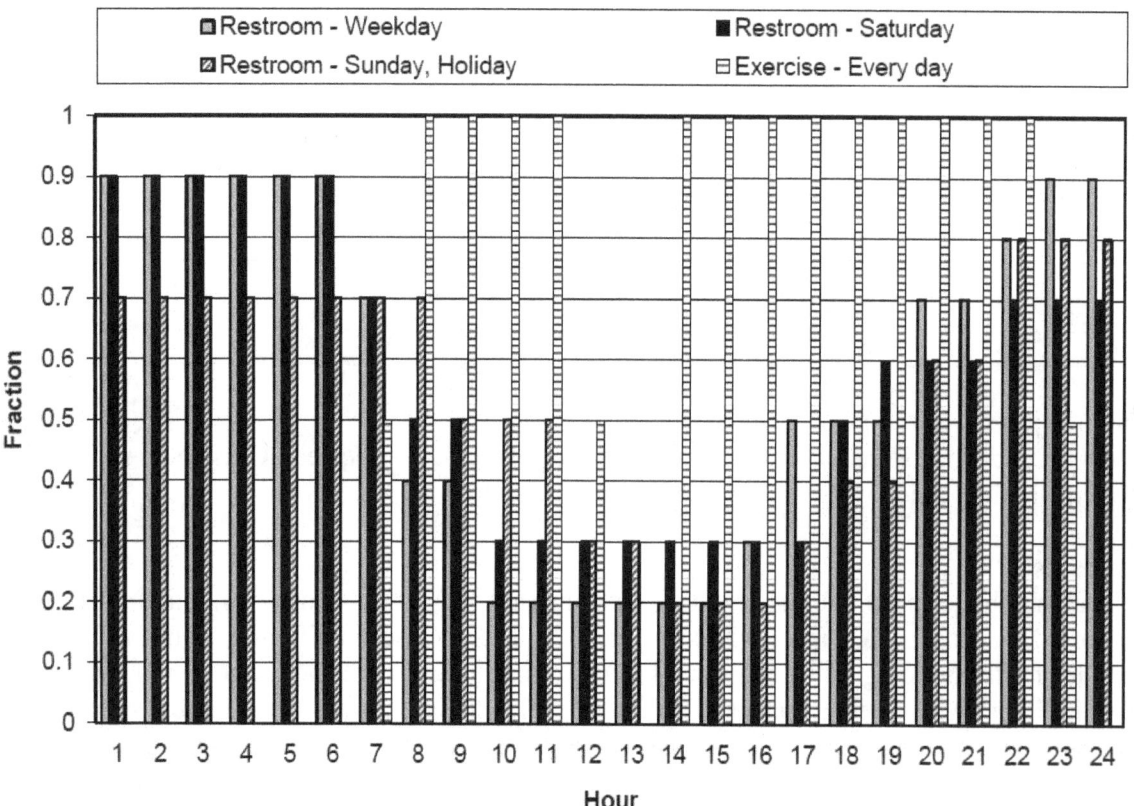

Figure A25 Occupancy schedules for Small Hotel (Restroom and Exercise)

Occupancy Schedules

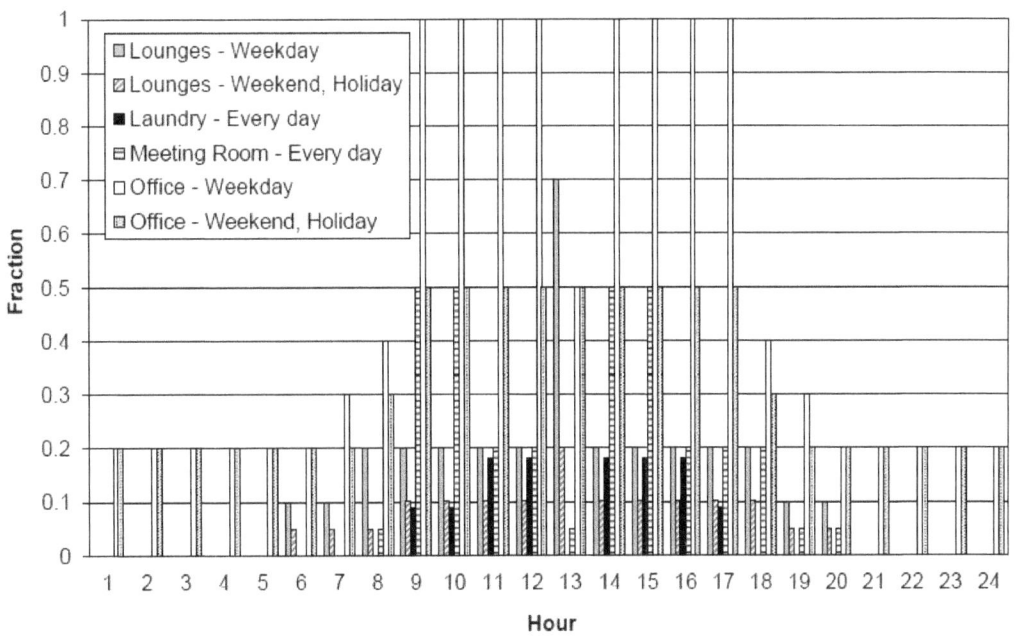

**Figure A26 Occupancy schedules for Small Hotel
(Lounges, Laundry, Meeting Room, Office)**

Occupancy Schedules

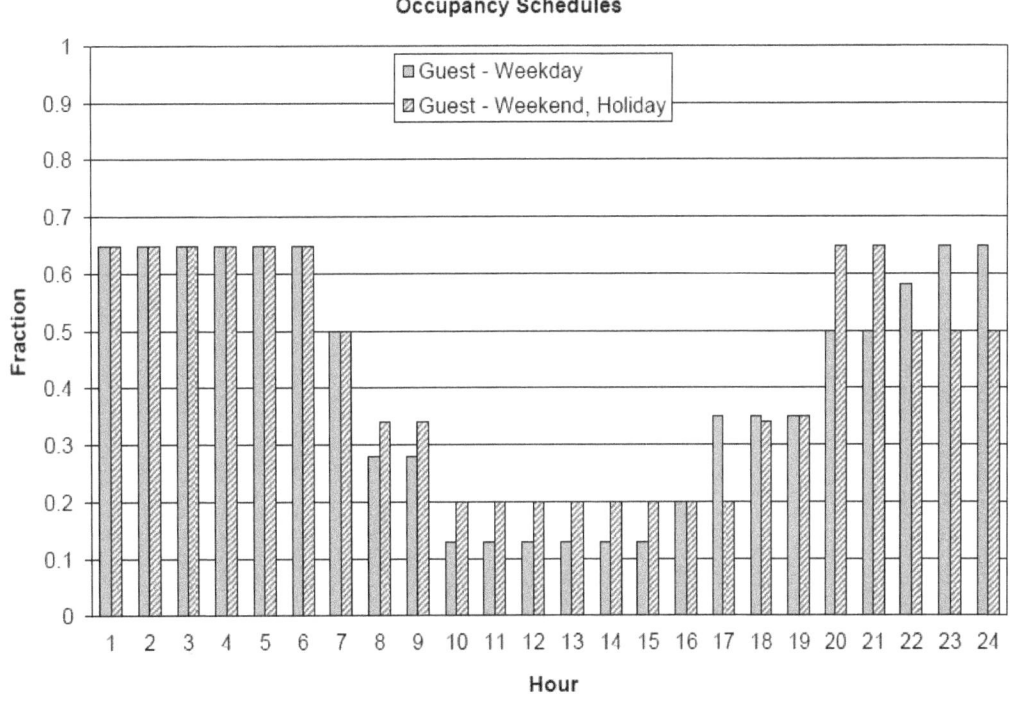

Figure A27 Occupancy schedule for Small Hotel (Guest)

A2.12. Large Hotel

Table A25 summarizes the zones modeled in CONTAM for the Large Hotel, their respective sizes, and maximum occupancy.

Table A25 Summary of zones in Large Hotel

Zone	Floor	Area (m^2)	Height (m)	Maximum occupancy
Basement	B	1919	2.44	53.25
Stair NW	B	20	2.44	0
Elevator NW	B	19	2.44	0
Stair SE	B	11	2.44	0
Elevator SE	B	10	2.44	0
Retail 1	1	67	3.96	10.83
Retail 2	1	78	3.96	12.54
Mechanical Room	1	164	3.96	0.00
Storage	1	95	3.96	2.04
Laundry	1	78	3.96	3.36
Cafe	1	189	3.96	135.52
Lobby	1	1239	3.96	422.48
Stair NW	1	20	3.96	0
Elevator NW	1	19	3.96	0
Stair SE	1	11	3.96	0
Elevator SE	1	10	3.96	0
Restroom	1	9	3.96	0
Room 1	2-5	39	3.05	1.50
Room 2	2-5	39	3.05	1.50
Room 3 (x19)	2-5	25	3.05	1.50
Room 4 (x19)	2-5	25	3.05	1.50
Room 5	2-5	39	3.05	1.50
Room 6	2-5	39	3.05	1.50
Corridor	2-5	329	3.05	4.19
Stair NW	2-5	20	3.05	0
Elevator NW	2-5	19	3.05	0
Stair SE	2-5	11	3.05	0
Elevator SE	2-5	10	3.05	0
Room 1	6	39	3.05	1.50
Room 2	6	39	3.05	1.50
Room 3 (x9)	6	25	3.05	1.50
Banquet	6	332	3.05	238.00
Dining	6	332	3.05	238.00
Kitchen	6	103	3.05	5.56
Corridor	6	311	3.05	4.44
Stair NW	6	20	3.05	0
Elevator NW	6	19	3.05	0
Stair SE	6	11	3.05	0
Elevator SE	6	10	3.05	0
Restroom	6	41	3.05	0

Geometry:

1979 m² footprint, six-story building (plus a basement) with flat roof. Total floor area is 11 345 m² which includes the Basement. The EnergyPlus model has seven zones on the first floor, 43 zones each on floors two to five, and 12 zones on the sixth floor. The upper floors (floors two to five) have identical floor plans. On the upper floors, there are zones that represent several rooms lumped together (Room 3 and Room 4 in Figure A28b, and Room 3 in Figure A28c). These are also modeled as single zones in CONTAM.

In CONTAM, the following zones are added (shaded in Figure A28):
- A 3.05 m × 3.05 m Restroom is added on the first floor between the Laundry and Storage zones, carved out of the Lobby;
- A 3.66 m × 11.28 m Restroom is added on the sixth floor across from the Dining zone, carved out of the Corridor;
- A Stairwell and Elevator Shaft are added near the northwest corner of the building, carved out of the Lobby or Corridor. The Stairwell is 3 m × 6.71 m, and the Elevator Shaft is 2.79 m × 6.71 m;
- A Stairwell and Elevator Shaft are added near the southeast corner of the building, carved out of the Lobby or Corridor. The Stairwell is 3 m × 3.66 m and the Elevator Shaft is 2.79 m × 3.66 m.

Large interior leakage paths were defined as follows:
- Between Retail 1 and Lobby on the first floor, and between Retail 2 and Lobby, zones, a single large leakage path of 4 m² is modeled;
- Between Café and Lobby zones on the first floor, a single large leakage path of 8 m² is modeled;
- Between Laundry and Restroom zones on the first floor, a single large leakage path of 4 m² is modeled;
- Between Laundry and Lobby zones on the first floor, and between Restroom and Lobby zones on the first floor, a 0.186 m² transfer grille is modeled;
- Between Restroom and Corridor zones on the sixth floor, two 0.186 m² transfer grilles are modeled;
- Between Kitchen and Dining zones on the sixth floor, a single large leakage path that is 50 % of the total wall area between the spaces is modeled;
- A stairwell is defined using CONTAM's stair shaft model for closed treads and zero people;
- An elevator shaft is defined using CONTAM's elevator shaft model.

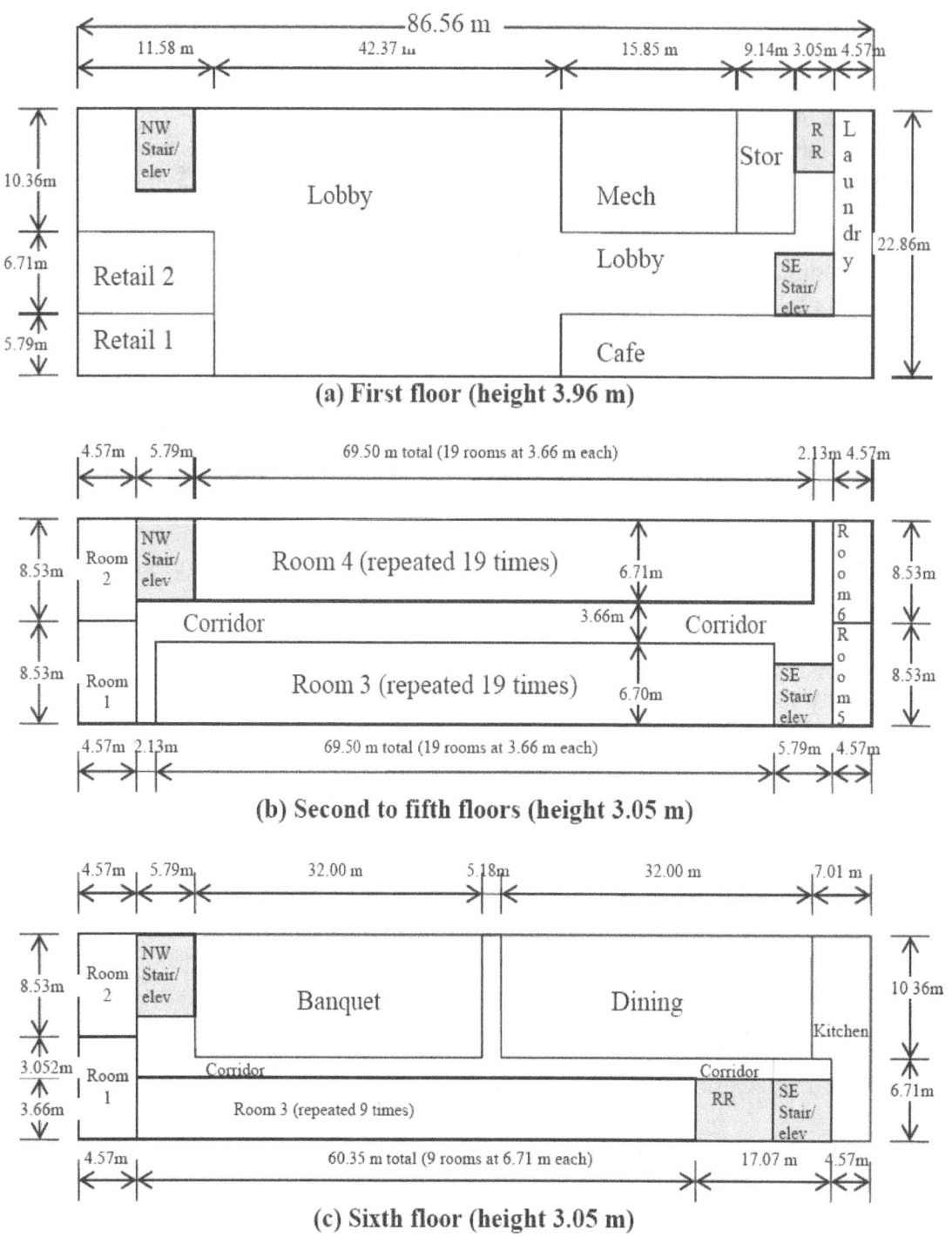

(a) First floor (height 3.96 m)

(b) Second to fifth floors (height 3.05 m)

(c) Sixth floor (height 3.05 m)

Figure A28 (a) First, (b) second to fifth, and (c) sixth floors plans of Large Hotel

HVAC systems:

For the New and Post-1980 buildings, the EnergyPlus model has one VAV system serving the common areas, including the Basement. The design supply flow rate calculated by EnergyPlus for each VAV system is used as the supply flow rate for the *constant*-volume system modeled in CONTAM for simplicity. The system modeled in CONTAM is still referred to as a "VAV" system in the body of this text. Varying the supply flow rate can be implemented in CONTAM using controls and/or schedules by users who wish to do so. The supply air, return air, outside air, and exhaust flow rates modeled in CONTAM are listed in Table A26 for the New and Post-1980 buildings. For the Pre-1980 buildings, the building has one CAV system serving the common areas including the Basement. Similarly, a constant-volume system is modeled in CONTAM. The supply air, return air, outside air, and exhaust flow rates modeled in CONTAM are listed in Table A27 for the Pre-1980 building. The exhaust flow rates for the Restrooms were modeled only in CONTAM, not in EnergyPlus.

In EnergyPlus, there was also a Dining exhaust fan (1.84 m^3/s) in addition to the Kitchen exhaust fan (0.04 m^3/s) and Laundry exhaust fan (0.24 m^3/s). The Dining exhaust fan was included in order to transfer air from the Dining zone to the Kitchen. This is modeled in CONTAM using a large opening between the Dining and Kitchen zones (see above), and one larger exhaust fan in the Kitchen (1.88 m^3/s). The sum of the Dining and Kitchen exhaust rates in EnergyPlus is 1.88 m^3/s.

For all building vintages, the EnergyPlus model has two dedicated outdoor air systems (DOAS) – one serves the guestrooms on the first floor and the other serves the guestrooms on floor two through five. Both systems deliver 100 % outside air. Similarly, two constant-volume systems are modeled in CONTAM. The outside air (also the supply air) and exhaust flow rates modeled in CONTAM are listed in Table A28. No air is returned to the DOAS in CONTAM.

In EnergyPlus, fan coil units in each guest room provide temperature control. They recirculate air locally within the zone and do not impact whole-building airflow or introduce outside air. In EnergyPlus, the sixth floor Corridor has a unit heater that recirculates air locally within the zone and does not impact whole-building airflow or introduce outside air. The fan coils and unit heater are not modeled in CONTAM.

In EnergyPlus, neutral building pressurization is modeled in all zones. To pressurize the building in CONTAM, less air is returned than is supplied to each zone. For all building vintages, in the common areas (Table A26 and Table A27), the return airflow rate is set to 90 % of the supply airflow rate. The return airflow from the Dining zone is reduced by the Kitchen exhaust (1.88 m^3/s) for all building vintages. For all building vintages, in the guestrooms (Table A28), the exhaust flow rates are 90 % of the outside air (or supply air) flow rates. The exhaust flow rates for the guestrooms were modeled only in CONTAM, not in EnergyPlus.

Table A26 Summary of VAV system flow rates (m³/s) in Large Hotel for New and Post-1980 buildings

Zone	Floor	New		Post-1980		Outside Air (m³/s)	Exhaust air (m³/s)
		Supply	Return	Supply	Return		
Basement	B	4.25	3.83	4.62	4.16	1.48	0
Retail 1	1	0.51	0.46	0.70	0.63	0.20	0
Retail 2	1	0.59	0.54	0.80	0.72	0.23	0
Mechanical Room	1	0.10	0.09	0.04	0.04	0.03	0
Storage	1	0.07	0.07	0.07	0.07	0.02	0
Laundry	1	5.54	4.75	5.61	4.81	1.86	0.24
Cafe	1	2.04	1.84	2.27	2.04	0.72	0
Lobby	1	9.24	8.28	10.40	9.33	3.27	0
Restroom	1	N/A	N/A	N/A	N/A	0	0.04
Corridor	2-5	0.10	0.09	0.34	0.30	0.07	0
Banquet	6	4.21	3.79	4.81	4.33	1.50	0
Dining	6	4.34	2.07	4.93	2.60	1.54	0
Kitchen	6	1.99	1.75	2.04	1.80	0.67	1.88
Corridor	6	0.42	0.17	0.45	0.19	0.14	0
Restroom	6	N/A	N/A	N/A	N/A	0	0.21
Total VAV		**33.72**	**27.98**	**38.08**	**31.91**	**11.94**	

Note: Exhaust air rates are the same for all building vintages.

Table A27 Summary of CAV system flow rates (m³/s) in Large Hotel for Pre-1980 buildings

Zone	Floor	Pre-1980		Outside Air (m³/s)
		Supply	Return	
Basement	B	3.80	3.42	1.22[1]
Retail 1	1	0.69	0.62	0.22
Retail 2	1	0.80	0.72	0.26
Mechanical Room	1	0.06	0.06	0.02
Storage	1	0.07	0.07	0.02
Laundry	1	5.58	4.78	1.79
Cafe	1	2.44	2.20	0.78
Lobby	1	10.28	9.22	3.30
Corridor	2-5	0.34	0.31	0.11[1]
Banquet	6	4.72	4.25	1.52
Dining	6	4.84	2.52	1.55
Kitchen	6	2.01	1.77	0.65
Corridor	6	0.54	0.28	0.17
Total CAV		**37.19**	**31.11**	**11.94**

Note 1: Compared with the outside air requirements for the New and Post-1980 buildings, the largest differences are for the Basement and Corridor zones in the Pre-1980 buildings.

110

Table A28 Summary of DOAS flow rates (m³/s) in Large Hotel for all building vintages

Zone	Floor	Outside Air (m³/s)	Exhaust air (m³/s)
Room 1	2-5	0.014	0.013
Room 2	2-5	0.014	0.013
Room 3 (x19)	2-5	0.27	0.24
Room 4 (x19)	2-5	0.27	0.24
Room 5	2-5	0.014	0.013
Room 6	2-5	0.014	0.013
DOAS 1		**2.35**	
Room 1	6	0.014	0.013
Room 2	6	0.014	0.013
Room 3 (x9)	6	0.13	0.11
DOAS 2		**0.15**	

Schedules:

The DOAS and exhaust fans operate 24 hours per day every day of the year. Outside air is also supplied all of the time. The VAV and CAV fans operate on the following schedule:

• Every day: on from 7:00 a.m. to 12:00 a.m., off otherwise

Outside air is supplied according to this schedule as well for the VAV and CAV systems.

Occupants:

The peak number of people for each zone is listed in Table A25. Occupants in all building zones are scheduled according to Figure A29 to Figure A30. There are three different occupant schedules for the building. The occupancy schedules for the guestrooms (referred to as "Guest" occupancy) and Lobby zone are shown in Figure A29. The occupancy schedule for the remaining zones is shown in Figure A30. For some zones, there is a different occupancy schedule for weekdays, Saturdays, and Sundays/ holidays.

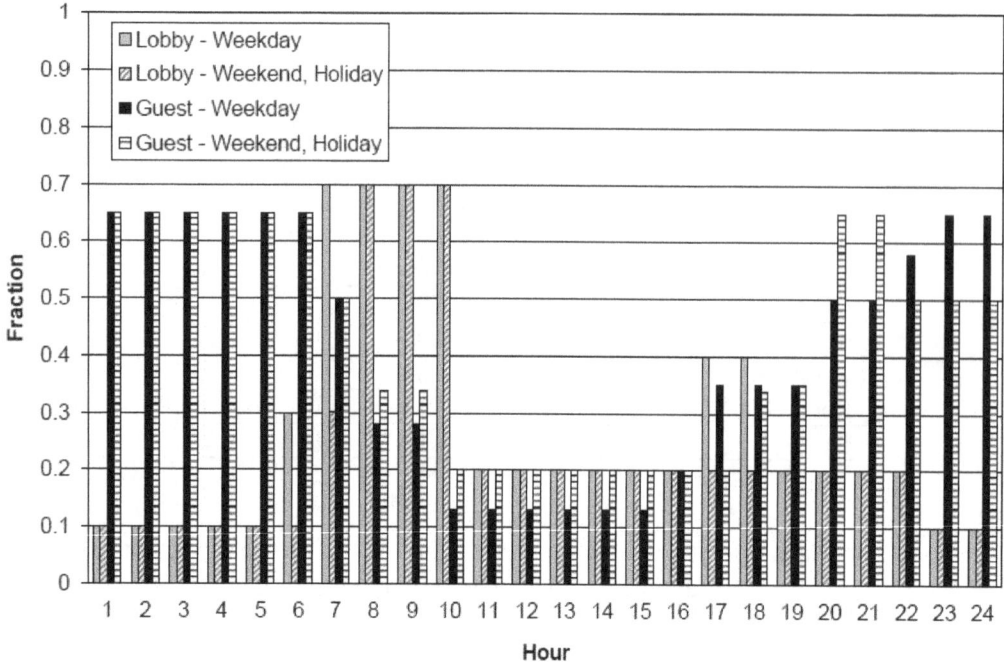

Figure A29 Occupancy schedules for Large Hotel (Lobby, Guest)

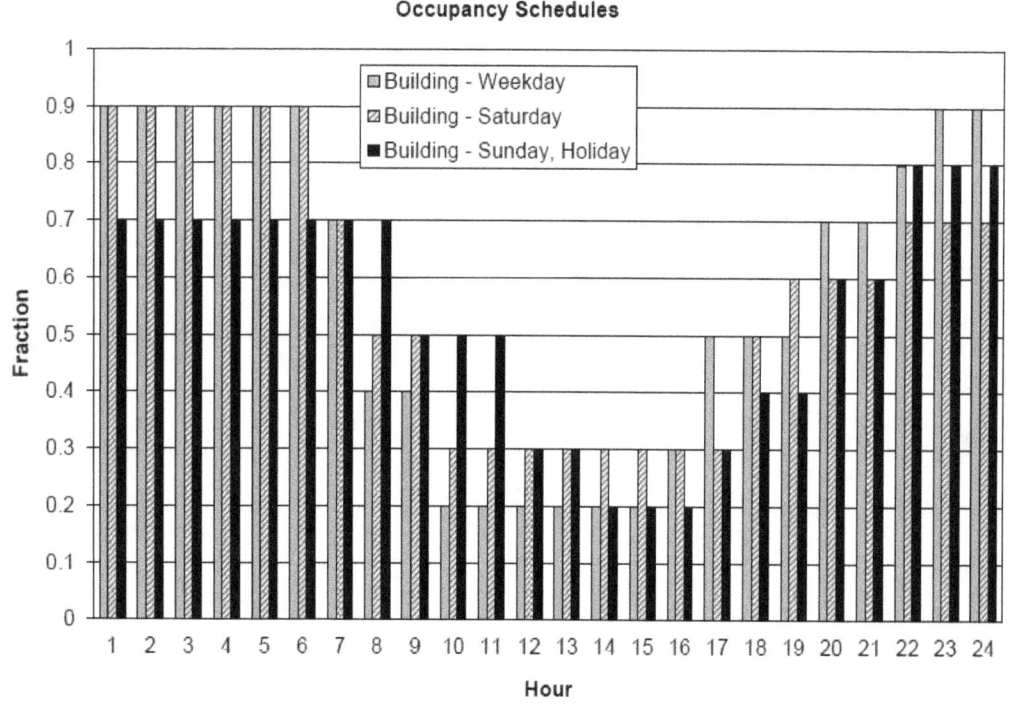

Figure A30 Occupancy schedule for Large Hotel (Building)

112

A2.13. Hospital

Table A29 summarizes the zones modeled in CONTAM for the Hospital, their respective sizes, and maximum occupancy. The table also lists the occupancy schedule for each zone (details given below on schedule type).

Geometry:

3739 m^2 footprint, five-story building (plus a basement) with flat roof. Total floor area is $22\,436$ m^2 which includes the Basement. The EnergyPlus model has nine zones on the first floor, 11 zones on second, 13 zones on third and fourth floors, and eight zones on the fifth floor (Figure A31 to Figure A34). The third and fourth floors have identical floor plans.

In CONTAM, the following zones are added (shaded in Figure A31 to Figure A34):

- A 4.57 m $\times$ 7.62 m Restroom is added on the first floor between the Office 1 and Corridor zones, carved out of the Lobby;
- A 6.10 m $\times$ 9.14 m Restroom is added on the second floor, carved out of the OR Nurse Station;
- A 6.10 m $\times$ 9.14 m Restroom is added on the third and fourth floors, carved out of the Nurse Station Lobby;
- A 4.57 m $\times$ 12.19 m Restroom is added on the sixth floor, carved out of the Nurse Station;
- A Stairwell and Elevator Shaft are added near the north and south sides the building, carved out of the Basement, Corridor, or Nurse Station on the sixth floor. Both the Stairwell and Elevator Shaft are 4.46 m $\times$ 5.39 m

There are patient rooms, exam rooms, operating rooms, and offices that are modeled in EnergyPlus using "multipliers". These are modeled as single zones in CONTAM, except for three sets where the multiplied zones were not adjacent to one another. These are modeled as follows:

- On the first floor, ER Exam 3 is multiplied by a factor of four in the EnergyPlus model, with two times on each side of the building. It is modeled in CONTAM as two zones (ERExam3$\times$2a and ERExam3$\times$2b – indicating that two rooms are grouped into a zone at each location);
- On the third and fourth floors, Patient Room 1 is multiplied by a factor of 10, with five on each side of the building. It is modeled in CONTAM as two zones (PatRm1a$\times$5 and PatRm1b$\times$5 – indicating that five rooms are grouped into a zone at each location);
- On the third and fourth floors, Patient Room 5 is multiplied by a factor of 10 in the EnergyPlus model, with five on each side of the building. It is modeled in CONTAM as two zones (PatRm5a$\times$5 and PatRm5b$\times$5 – indicating that five rooms are grouped into a zone at each location).

Large interior leakage paths were defined as follows:

- Between Lobby and Corridor zones on the first floor, a single large leakage path that is 50 % of the total wall area between the spaces is modeled;
- Between ER Nurse Station and Corridor zones on the first floor, a single large leakage path of 8 m^2 is modeled;

Table A29 Summary of zones in Hospital

Zone	Floor	Area (m^2)	Height (m)	Maximum occupancy	Occupancy schedule (see below for details)
Basement	B	3643	2.44	100.6	Admin
ER_Exam1_Mult4	1	28	4.27	6.00	Critical
ER_Trauma1	1	28	4.27	6.00	Critical
ER_Exam3_Mult4	1	28	4.27	6.00	Critical
ER_Trauma2	1	28	4.27	6.00	Critical
ER_Triage_Mult4	1	28	4.27	6.00	Critical
Office1_Mult4	1	14	4.27	1.05	Admin
Lobby_Records	1	1440	4.27	21.09	Admin
Corridor	1	473	4.27	6.15	Admin
ER_NurseStn_Lobby	1	1236	4.27	17.67	Admin
OR1	2	56	4.27	3.00	Critical
OR2_Mult5	2	56	4.27	3.00	Critical
OR3	2	56	4.27	3.00	Critical
OR4	2	223	4.27	12.00	Critical
IC_PatRoom1_Mult5	2	21	4.27	1.12	Critical
IC_PatRoom2	2	28	4.27	1.50	Critical
IC_PatRoom3_Mult6	2	21	4.27	1.12	Critical
ICU	2	618	4.27	33.25	Critical
ICU_NurseStnLobby	2	669	4.27	9.56	Admin
Corridor	2	473	4.27	6.15	Admin
OR_NurseStn_Lobby	2	957	4.27	14.48	Admin
PatRoom1_Mult10	3	21	4.27	1.12	Critical
PatRoom2	3	35	4.27	1.88	Critical
PatRoom3_Mult10	3	20	4.27	1.09	Critical
PatRoom4	3	35	4.27	1.88	Critical
PatRoom5_Mult10	3	21	4.27	1.12	Critical
PhysTherapy	3	488	4.27	26.25	Admin
PatRoom6	3	28	4.27	1.50	Critical
PatRoom7_Mult10	3	20	4.27	1.09	Critical
PatRoom8	3	28	4.27	1.50	Critical
NurseStn_Lobby	3	850	4.27	12.95	Admin
Lab	3	265	4.27	14.25	Admin
Corridor_SE	3	518	4.27	6.12	Admin
Corridor_NW	3	518	4.27	6.12	Admin
PatRoom1_Mult10	4	21	4.27	1.12	Critical
PatRoom2	4	35	4.27	1.88	Critical
PatRoom3_Mult10	4	20	4.27	1.09	Critical
PatRoom4	4	35	4.27	1.88	Critical
PatRoom5_Mult10	4	21	4.27	1.12	Critical
Radiology	4	488	4.27	26.25	Admin
PatRoom6	4	28	4.27	1.50	Critical
PatRoom7_Mult10	4	20	4.27	1.09	Critical
PatRoom8	4	28	4.27	1.50	Critical
NurseStn_Lobby	4	850	4.27	12.95	Admin
Lab	4	265	4.27	14.25	Admin

Zone	Floor	Area (m²)	Height (m)	Maximum occupancy	Occupancy schedule (see below for details)
Corridor_SE	4	518	4.27	6.12	Admin
Corridor_NW	4	518	4.27	6.12	Admin
Dining	5	641	4.27	75.00	Admin
NurseStn_Lobby	5	993	4.27	14.88	Admin
Kitchen	5	929	4.27	50.00	Admin
Office1	5	70	4.27	5.25	Admin
Office2_Mult5	5	70	4.27	5.25	Admin
Office3	5	70	4.27	5.25	Admin
Office4_Mult6	5	14	4.27	1.05	Admin
Corridor	5	454	4.27	5.42	Admin
Restroom	1	35	4.27	0	N/A
Restroom	2	56	4.27	0	N/A
Restroom	3	56	4.27	0	N/A
Restroom	4	56	4.27	0	N/A
Restroom	5	56	4.27	0	N/A
Stair 1	all	24	4.27	0	N/A
Stair 2	all	24	4.27	0	N/A
Elev 1	all	24	4.27	0	N/A
Elev2	all	24	4.27	0	N/A

- Between ICU Nurse Station and OR Nurse Station zones on the second floor, a single large leakage path of 8 m² is modeled;
- Between Nurse Station Lobby and Corridors zones on the fifth floor, two large leakage paths of 4 m² are modeled;
- Between Nurse Station Lobby and Dining zones on the sixth floor, one single large leakage path of 4 m² is modeled;
- All Restrooms are modeled with a 0.186 m² transfer grille in the door between the zone and adjacent zone (Corridor for floors one to four, Dining for sixth floor).
- A stairwell is defined using CONTAM's stair shaft model for closed treads and zero people;
- An elevator shaft is defined using CONTAM's elevator shaft model.

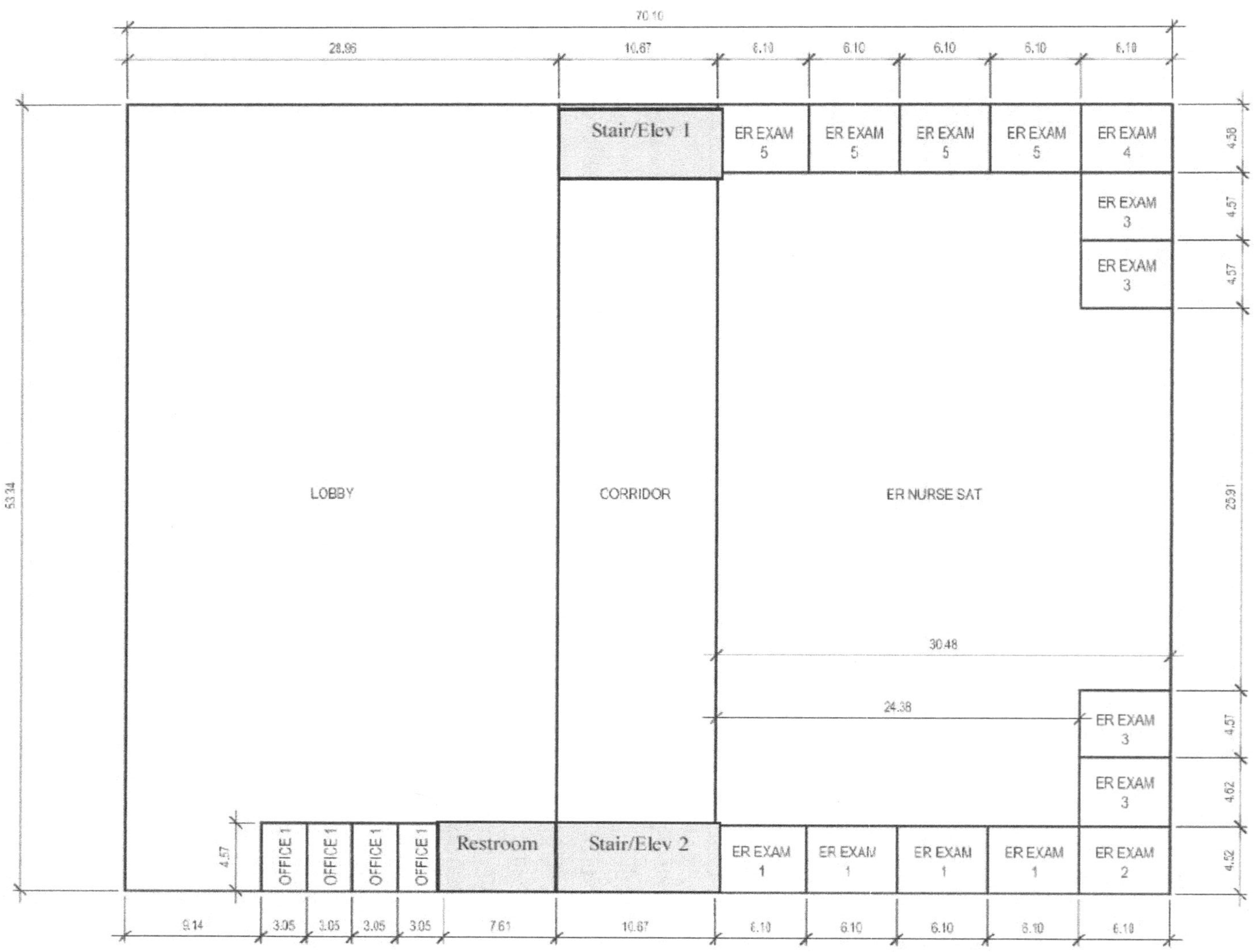

Figure A31 First floor plan of Hospital (height 4.27 m), all dimensions in meters

116

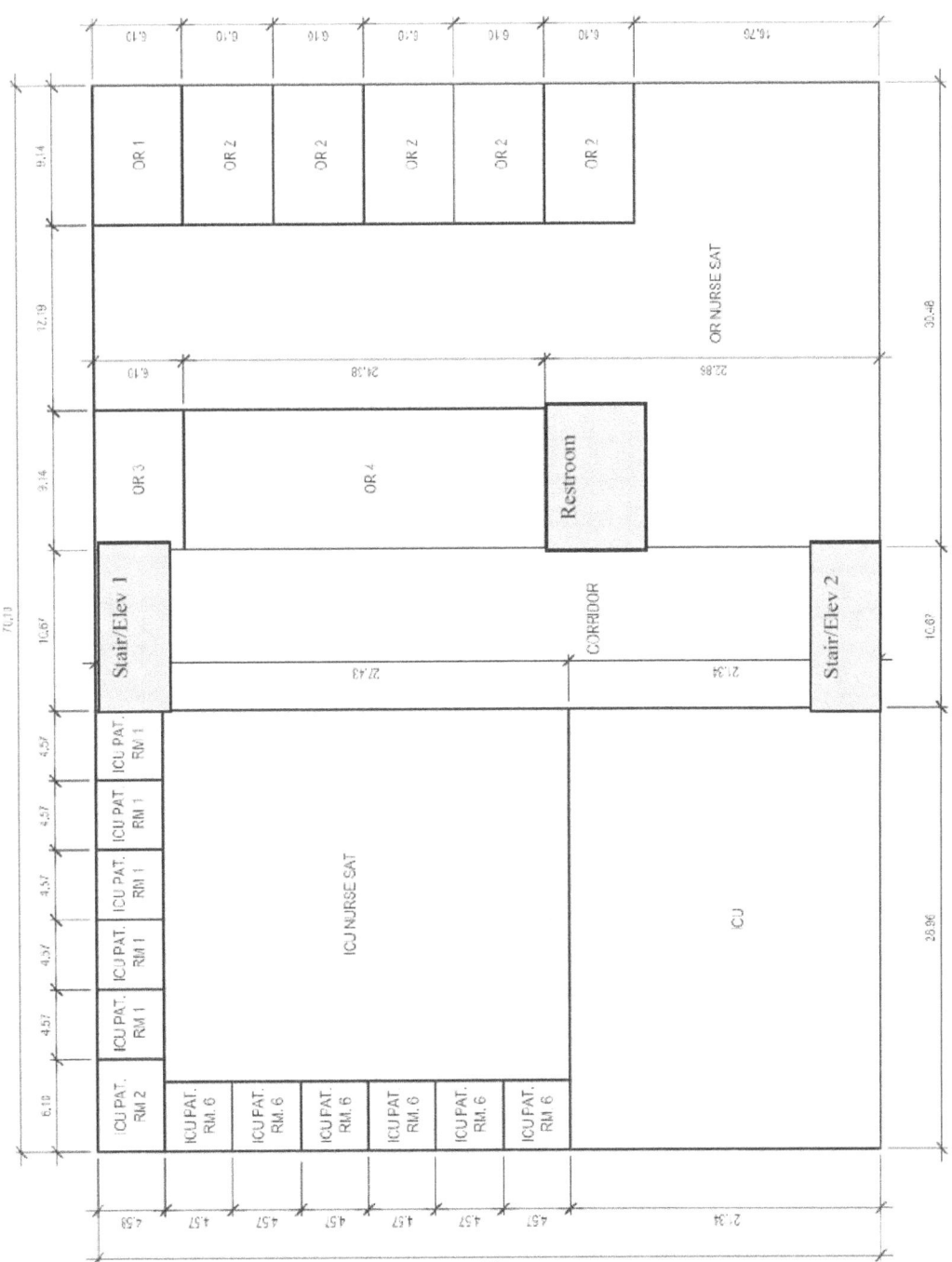

Figure A32 Second floor plan of Hospital (height 4.27 m), all dimensions in meters

117

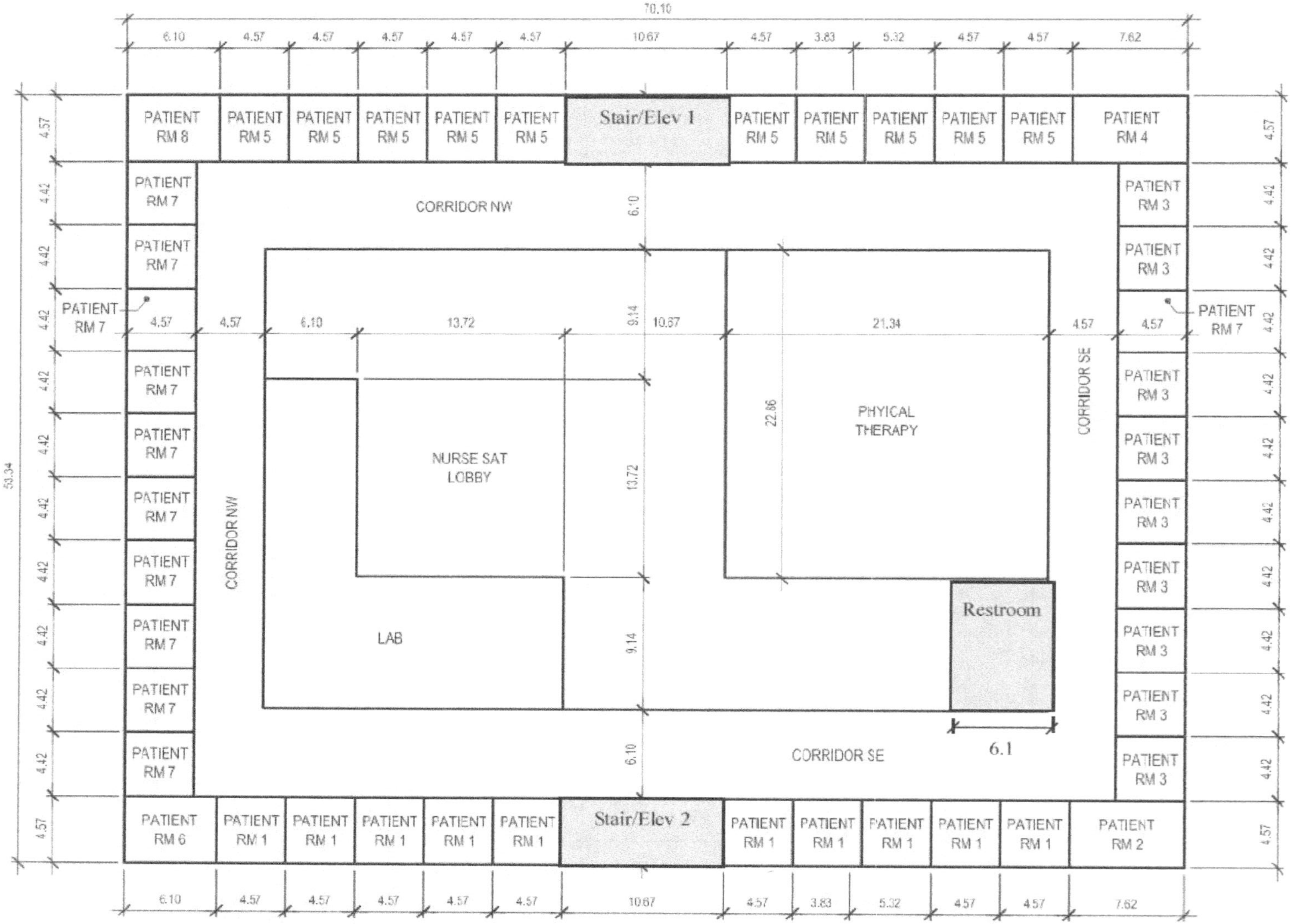

Figure A33 Third/Fourth floor plans of Hospital (height 4.27 m), all dimensions in meters

118

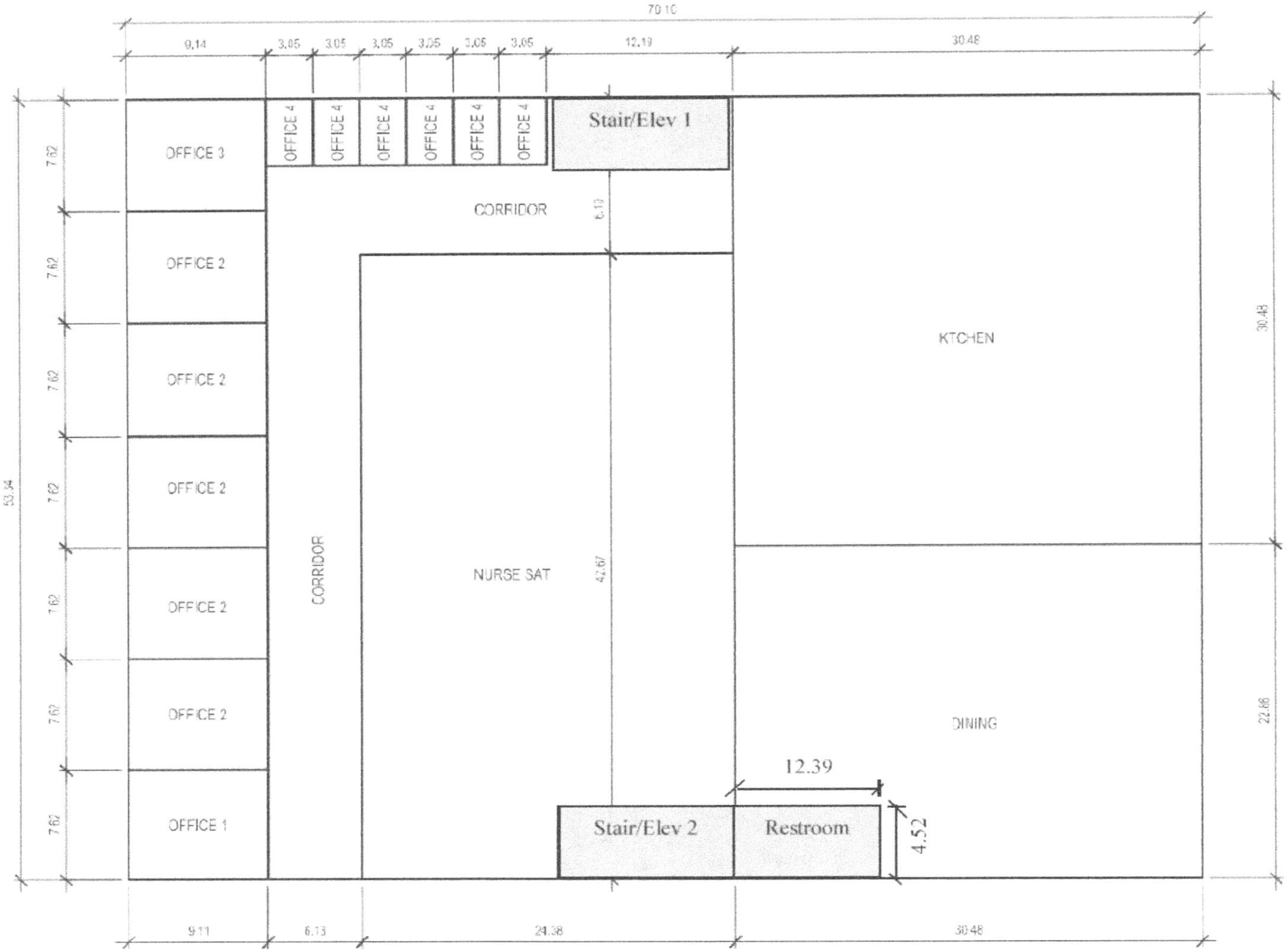

Figure A34 Fifth floor plans of Hospital (height 4.27 m), all dimensions in meters

119

HVAC systems:

For all building vintages, the EnergyPlus model has two VAV systems and two CAV systems. The two VAV systems serve the "administrative" zones, and the two CAV systems serve the "critical" zones. The design supply flow rate calculated by EnergyPlus for each VAV system is used as the supply flow rate for the *constant*-volume system modeled in CONTAM for simplicity. The system modeled in CONTAM is still referred to as a "VAV" system in the body of this text. Varying the supply flow rate can be implemented in CONTAM using controls and/or schedules by users who wish to do so. The supply air, return air, outside air, and exhaust flow rates modeled in CONTAM are listed in Table A30 for both the VAV and CAV systems. In EnergyPlus, each HVAC system is assigned a minimum ventilation requirement and is modeled as such in CONTAM. For CAV 1, the minimum ventilation requirement is 4.16 m^3/s, it is 2.54 m^3/s for CAV 2, it is 5.18 m^3/s for VAV 1, and it is 9.49 m^3/s for VAV 2 for all building vintages. The exhaust flow rates for the Restrooms were modeled only in CONTAM, not in EnergyPlus.

In EnergyPlus, there was also a Dining exhaust fan (0.75 m^3/s) in addition to the Kitchen exhaust fan (1.75 m^3/s). The Dining exhaust fan was included in order to transfer air from the Dining zone to the Kitchen. This is modeled in CONTAM using a large opening between the Dining and Kitchen zones (see above), and one larger exhaust fan in the Kitchen (2.5 m^3/s). The sum of the Dining and Kitchen exhaust rates in EnergyPlus is 2.5 m^3/s.

In EnergyPlus, neutral building pressurization is modeled in all zones. To pressurize the building in CONTAM, less air is returned than is supplied to each zone. For all building vintages, the return airflow rate is set to 90 % of the supply airflow rate. The return airflow from the Kitchen is reduced by 1.75 m^3/s, and the return airflow from the Dining is reduced by 0.75 m^3/s, to allow makeup air for the Kitchen exhaust (2.5 m^3/s).

Schedules:

All the HVAC system and exhaust fans operate 24 hours per day every day of the year. Outside air is also supplied all of the time.

Occupants:

The peak number of people for each zone is listed in Table A29. Occupants in all building zones are scheduled according to Figure A35. There are two different occupant schedules for the building. The occupancy schedule for the critical is for exam rooms, trauma rooms, triage, operating rooms, and patient rooms. The "Admin" occupancy schedule is for the remaining zones.

Table A30 Summary of VAV and CAV system flow rates (m³/s) in Hospital

Zone	Floor	New		Post-1980		Pre-1980		Outside Air (m³/s)	Exhaust air (m³/s)
		Supply	Return	Supply	Return	Supply	Return		
ER_Exam1_Mult4	1	1.59	1.43	1.59	1.43	1.59	1.43		0
ER_Exam3_Mult4	1	1.59	1.43	1.59	1.43	1.59	1.43		0
ER_Trauma1	1	0.40	0.36	0.40	0.36	0.40	0.36		0
ER_Trauma2	1	0.40	0.36	0.40	0.36	0.40	0.36		0
ER_Triage_Mult4	1	1.59	1.43	1.59	1.43	1.59	1.43		0
IC_PatRoom1_Mult5	2	0.74	0.67	0.86	0.77	0.87	0.79		0
PatRoom1_Mult10	3	1.49	1.34	1.49	1.34	1.49	1.34		0
PatRoom5_Mult10	3	1.49	1.34	1.49	1.34	1.49	1.34		0
PatRoom7_Mult10	3	1.67	1.50	2.03	1.83	2.14	1.93		0
PatRoom3_Mult10	4	1.44	1.29	1.60	1.44	1.68	1.51		0
PatRoom5_Mult10	4	1.49	1.34	1.49	1.34	1.49	1.34		0
PatRoom6	4	0.23	0.21	0.29	0.26	0.31	0.28		0
PatRoom7_Mult10	4	1.66	1.50	2.02	1.82	2.13	1.92		0
CAV 1 Total		**15.76**	**14.18**	**16.83**	**15.15**	**17.16**	**15.45**	**4.16**	
IC_PatRoom2	2	0.25	0.22	0.34	0.31	0.36	0.32		0
OR1	2	0.99	0.89	1.36	1.22	1.37	1.23		0
OR2_Mult5	2	4.96	4.46	6.66	5.99	6.69	6.02		0
OR3	2	0.99	0.89	1.33	1.20	1.33	1.20		0
OR4	2	3.97	3.57	5.00	4.50	5.00	4.50		0
PatRoom2	3	0.25	0.22	0.28	0.25	0.30	0.27		0
PatRoom6	3	0.23	0.21	0.29	0.26	0.31	0.28		0
PatRoom8	3	0.22	0.20	0.28	0.25	0.30	0.27		0
PatRoom4	4	0.25	0.22	0.26	0.24	0.27	0.25		0
CAV 2 Total		**12.10**	**10.89**	**15.81**	**14.23**	**15.92**	**14.33**	**2.54**	
Basement	B	4.18	3.76	5.64	5.08	5.39	4.85		0
Corridor	1	0.84	0.76	0.96	0.87	1.01	0.91		0
ER_NurseStn_Lobby	1	3.62	3.25	4.10	3.69	4.13	3.72		0
Lobby_Records	1	3.74	3.37	3.96	3.56	4.15	3.74		0
Office1_Mult4	1	0.25	0.23	0.34	0.31	0.37	0.33		0
Corridor	2	1.35	1.21	1.35	1.22	1.35	1.22		0
IC_PatRoom3_Mult6	2	1.16	1.05	1.55	1.40	1.62	1.46		0
ICU_NurseStn_Lobby	2	2.19	1.97	2.46	2.21	2.46	2.21		0
OR_NurseStn_Lobby	2	2.64	2.38	2.97	2.67	3.03	2.73		0
Lab	3	1.88	1.69	1.88	1.69	1.88	1.69		0
PatRoom3_Mult10	3	1.44	1.29	1.61	1.45	1.68	1.51		0
PatRoom1_Mult10	4	1.49	1.34	1.49	1.34	1.49	1.34		0
PatRoom8	4	0.22	0.20	0.28	0.25	0.29	0.27		0
VAV 1 Total		**25.00**	**22.50**	**28.60**	**25.74**	**28.85**	**25.97**	**5.18**	
ICU	2	4.40	3.96	4.63	4.17	4.71	4.24		0
Corridor_NW	3	1.34	1.21	1.34	1.21	1.34	1.21		0
Corridor_SE	3	1.34	1.21	1.34	1.21	1.34	1.21		0
NurseStn_Lobby	3	2.20	1.98	2.43	2.19	2.43	2.19		0
PatRoom4	3	0.25	0.22	0.26	0.24	0.28	0.25		0
PhysTherapy	3	3.47	3.12	3.47	3.12	3.47	3.12		0
Corridor_NW	4	1.34	1.21	1.34	1.21	1.34	1.21		0

121

Zone	Floor	New		Post-1980		Pre-1980		Outside	Exhaust
Corridor_SE	4	1.34	1.21	1.34	1.21	1.34	1.21		0
Lab	4	1.88	1.69	1.88	1.69	1.88	1.69		0
NurseStn_Lobby	4	2.25	2.02	2.48	2.23	2.48	2.23		0
PatRoom2	4	0.25	0.22	0.28	0.25	0.30	0.27		0
Radiology	4	8.67	7.81	8.67	7.80	8.67	7.80		0
Corridor	5	1.12	1.01	1.22	1.10	1.37	1.23		0
Dining	5	2.76	1.73	3.62	2.51	3.97	2.82		0
Kitchen	5	3.26	1.18	3.31	1.23	3.26	1.18		2.5
NurseStn_Lobby	5	3.23	2.90	3.32	2.99	3.73	3.36		0
Office1	5	0.49	0.44	0.58	0.53	0.64	0.58		0
Office2_Mult5	5	2.14	1.92	2.52	2.27	2.75	2.48		0
Office3	5	0.47	0.43	0.57	0.51	0.62	0.56		0
Office4_Mult6	5	0.38	0.35	0.50	0.45	0.56	0.50		0
VAV 2 Total		42.58	35.82	45.11	38.10	46.47	39.32	9.49	

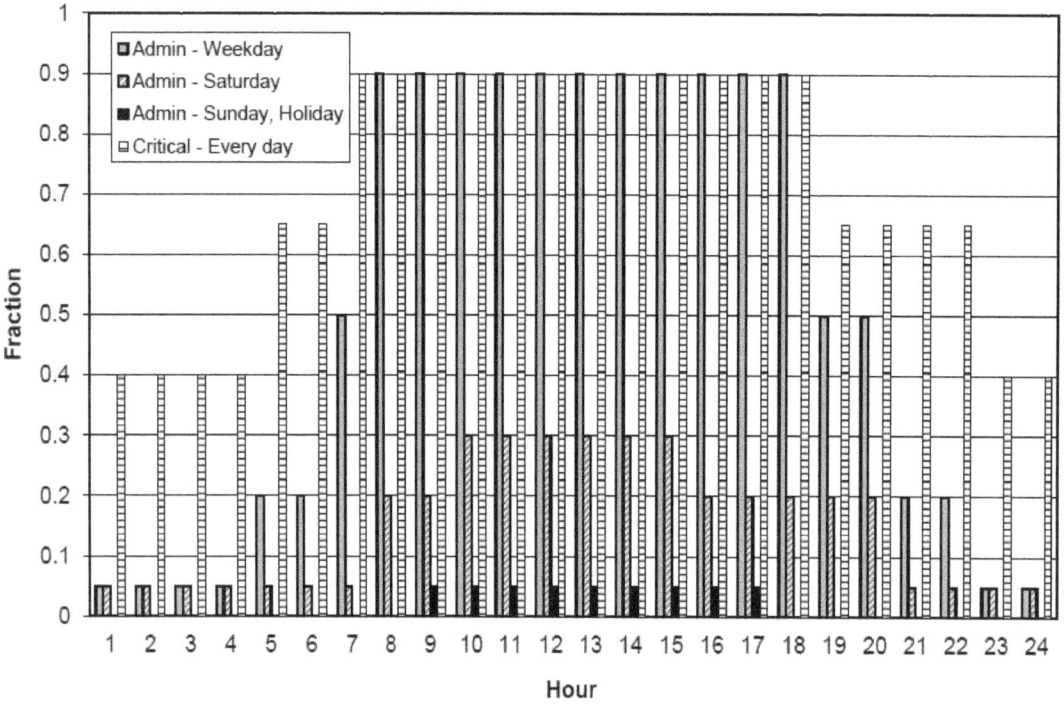

Figure A35 Occupancy schedules for Hospital

A2.14. Outpatient Health Care

Table A31 summarizes the zones modeled in CONTAM for the Outpatient Health Care, their respective sizes, and maximum occupancy.

Table A31 Summary of zones in Outpatient Health Care

Zone	Floor	Area (m^2)	Height (m)	Maximum occupancy
Anesthesia	1	10	3.05	0.00
Bio Hazard	1	5	3.05	0.00
Café	1	39	3.05	39.02
Clean	1	12	3.05	2.34
Clean Work	1	15	3.05	3.07
Dictation	1	12	3.05	0.59
Dressing Room	1	4	3.05	0.21
Electrical Room	1	9	3.05	0.00
Elevator Pump Room	1	8	3.05	0.00
Humid	1	5	3.05	0.25
IT Hall	1	13	3.05	0.00
IT Room	1	10	3.05	0.52
Lobby	1	58	3.05	17.34
Lobby Hall	1	22	3.05	0.00
Lobby Toilet	1	5	3.05	0.00
Locker Room	1	61	3.05	9.20
Locker Room Hall	1	46	3.05	0.00
Lounge	1	33	3.05	5.02
Med Gas	1	5	3.05	0.00
MRI Control Room	1	16	3.05	3.12
MRI Hall	1	14	3.05	0.00
MRI Room	1	41	3.05	8.18
MRI Toilet	1	5	3.05	0.00
Nourishment	1	17	3.05	3.38
Nurse Hall	1	46	3.05	0.00
Nurse Janitor	1	5	3.05	0.00
Nurse Station	1	24	3.05	4.85
Nurse Toilet	1	5	3.05	0.00
Office	1	45	3.05	2.24
Operating Room 1	1	43	3.05	8.55
Operating Room 2	1	45	3.05	8.92
Operating Room 3	1	44	3.05	8.84
PACU	1	10	3.05	2.01
Pre-Op Hall	1	49	3.05	0.00
Pre-Op Room 1	1	18	3.05	1.76
Pre-Op Room 2	1	31	3.05	3.14
Pre-Op Toilet	1	5	3.05	0.00
Procedure Room	1	26	3.05	5.30
Reception	1	47	3.05	14.19
Reception Hall	1	12	3.05	0.00
Recovery Room	1	50	3.05	10.03
Scheduling	1	11	3.05	0.55

Zone	Floor	Area (m²)	Height (m)	Maximum occupancy
Scrub	1	8	3.05	0.00
Soil	1	12	3.05	2.34
Soil Hold	1	5	3.05	1.04
Soil Work	1	17	3.05	3.34
Step Down	1	28	3.05	5.57
Sterile Hall	1	57	3.05	0.00
Sterile Storage	1	37	3.05	0.00
Storage	1	85	3.05	0.00
Sub-Sterile	1	18	3.05	0.00
Utility Hall	1	24	3.05	0.00
Utility Janitor	1	4	3.05	0.00
Utility Room	1	33	3.05	0.00
Vestibule	1	7	3.05	0.00
Conference	2	31	3.05	15.61
Conference Toilet	2	6	3.05	0.00
Dictation	2	7	3.05	0.33
Exam 1	2	33	3.05	6.69
Exam 2	2	50	3.05	10.03
Exam 3	2	67	3.05	13.38
Exam 4	2	8	3.05	1.56
Exam 5	2	33	3.05	6.50
Exam 6	2	21	3.05	4.18
Exam 7	2	74	3.05	14.72
Exam 8	2	25	3.05	5.02
Exam 9	2	37	3.05	7.36
Exam Hall 1	2	17	3.05	0.00
Exam Hall 2	2	17	3.05	0.00
Exam Hall 3	2	17	3.05	0.00
Exam Hall 4	2	18	3.05	0.00
Exam Hall 5	2	18	3.05	0.00
Exam Hall 6	2	18	3.05	0.00
Janitor	2	6	3.05	0.00
Lounge	2	7	3.05	1.11
Nurse Station 1	2	14	3.05	2.79
Nurse Station 2	2	17	3.05	3.34
Office	2	52	3.05	2.60
Office Hall	2	41	3.05	0.00
Reception	2	91	3.05	27.42
Reception Hall	2	52	3.05	0.00
Reception Toilet	2	12	3.05	0.00
Scheduling 1	2	30	3.05	1.51
Scheduling 2	2	32	3.05	1.59
Storage 1	2	5	3.05	0.00
Storage 2	2	11	3.05	0.00
Storage 3	2	13	3.05	0.00
Utility	2	12	3.05	0.00
Work	2	157	3.05	7.85

Zone	Floor	Area (m²)	Height (m)	Maximum occupancy
Work Hall	2	77	3.05	0.00
Work Toilet	2	5	3.05	0.00
X-Ray	2	84	3.05	16.72
Dressing Room	3	4	3.05	0.20
Elevator Hall	3	34	3.05	0.00
Humid	3	10	3.05	0.50
Janitor	3	6	3.05	0.00
Locker	3	11	3.05	1.67
Lounge	3	71	3.05	10.58
Lounge Toilet	3	18	3.05	0.00
Mechanical	3	33	3.05	0.00
Mechanical Hall	3	28	3.05	0.00
Office	3	282	3.05	14.10
Office Hall	3	77	3.05	0.00
Office Toilet	3	5	3.05	0.00
Physical Therapy 1	3	121	3.05	24.15
Physical Therapy 2	3	55	3.05	11.00
Physical Therapy Toilet	3	8	3.05	0.00
Storage 1	3	10	3.05	0.00
Storage 2	3	8	3.05	0.00
Treatment	3	44	3.05	8.84
Undeveloped 1	3	211	3.05	0.00
Undeveloped 2	3	107	3.05	0.00
Utility	3	20	3.05	0.00
Work	3	53	3.05	2.67
NE Stair	all	16	9.14	0.00
NW Elevator	all	13	9.14	0.00
NW Stair	all	18	9.14	0.00
SW Stair	all	9	9.14	0.00

Geometry:

1373 m² footprint, three-story building with flat roof. The second and third floors have a slightly smaller footprint of 1232 m² each. Total floor area is 3837 m². The EnergyPlus model has 55 zones on the first floor, 37 zones on second floor, and 22 zones the third floor. Three Stairwells and one Elevator Shaft are modeled in EnergyPlus as single tall zones with heights equal to the height of the building. In CONTAM, they are modeled as one zone per floor, with large leakage paths connecting them (see below).

Large interior leakage paths were defined as follows:
- All rooms with exhaust fans (see Table A31) are modeled with a 0.186 m² transfer grille in the door between the zone and adjacent zone. The exception is the MRI room which does not have a transfer grille modeled.
- Between the following zones a single large leakage path of 4 m² is modeled:
    - First floor: Between Lobby Hall and Lobby
    - First floor: Between Lobby Hall and Reception
    - First floor: Between Reception Hall and Pre-Op Hall

    □ First floor: Between Nurse Station and Nurse Hall

    □ First floor: Between Locker Room Hall and Pre-Op Hall

    □ First floor: Between Nurse Hall and Pre-Op Hall

    □ Second floor: Between Reception and Work Hall

    □ Second floor: Between Reception and Reception Hall

    □ Second floor: Between Office Hall and Work Hall

    □ Third floor: Between Elevator Hall and Office Hall

    □ Third floor: Between Mechanical Hall and Office Hall

- A stairwell is defined using CONTAM's stair shaft model for closed treads and zero people;
- An elevator shaft is defined using CONTAM's elevator shaft model.

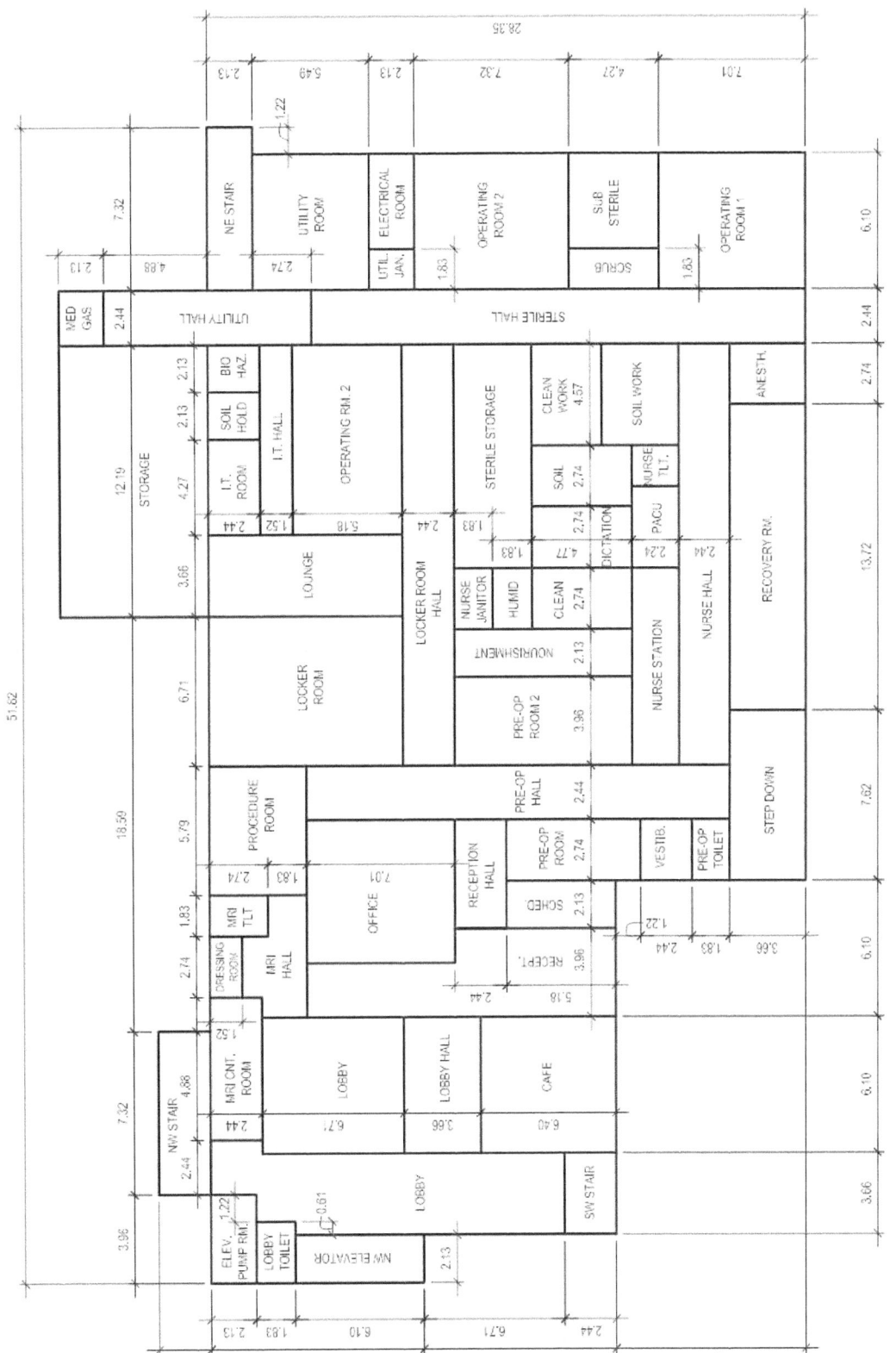

Figure A36 First floor plan of Outpatient Health Care, all dimensions in meters

127

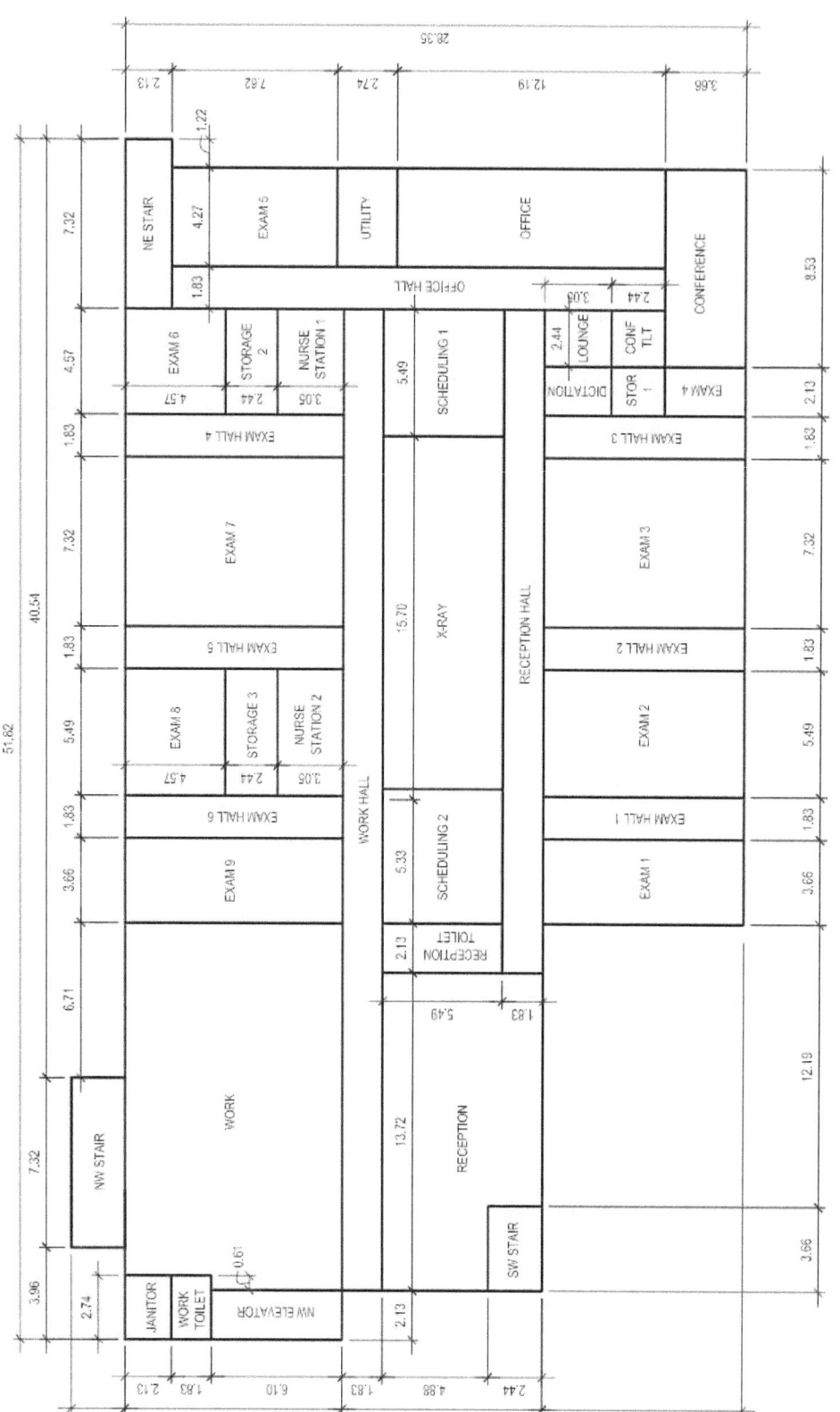

Figure A37 Second floor plan of Outpatient Health Care, all dimensions in meters

128

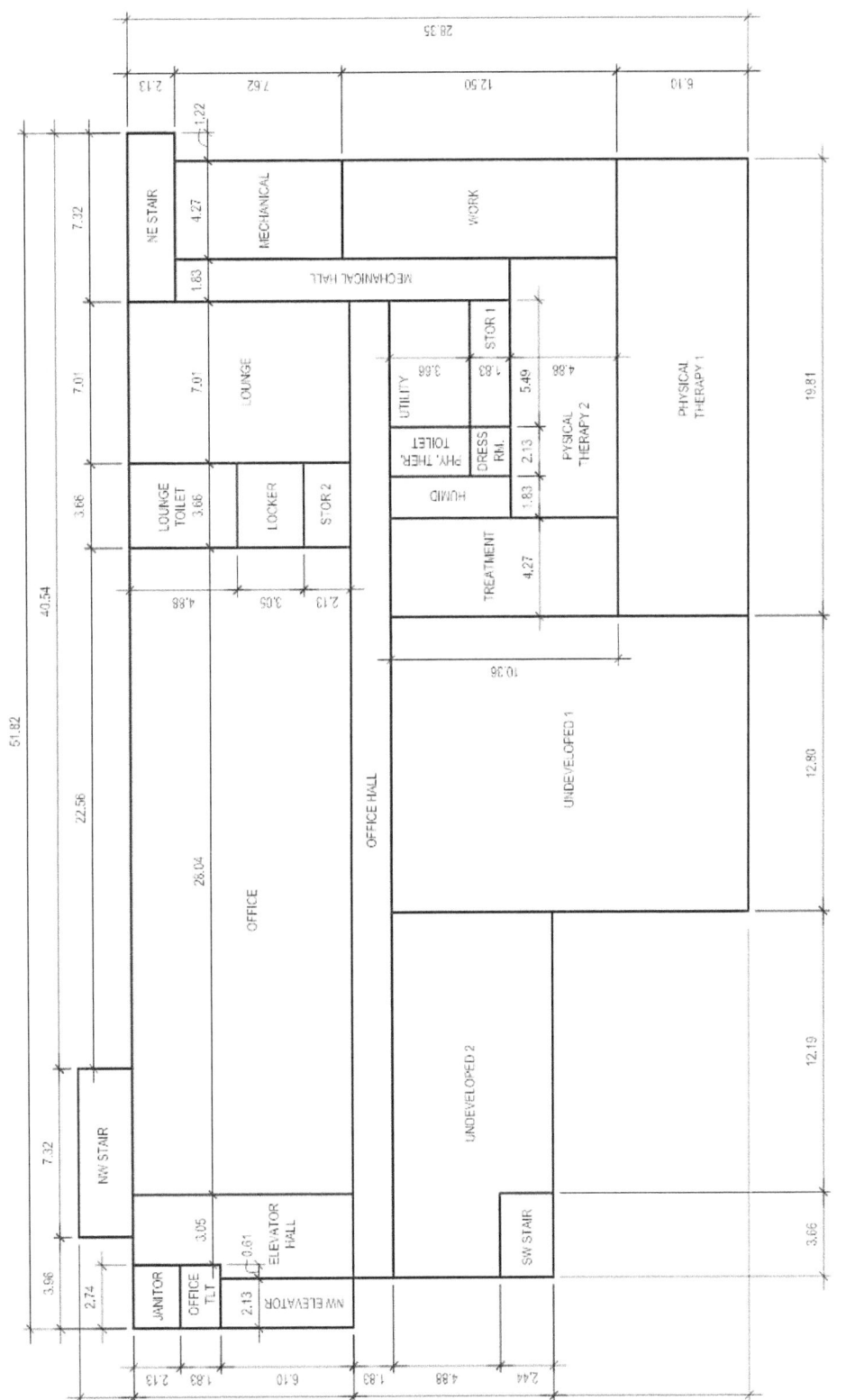

Figure A38 Third floor plan of Outpatient Health Care, all dimensions in meters

129

HVAC systems:

For all building vintages, the EnergyPlus model has two VAV (referred to as "AHU" systems here and in the models). One serves the first floor and the other serves the second and third floors. The design supply flow rate calculated by EnergyPlus for each VAV system is used as the supply flow rate for the *constant*-volume system modeled in CONTAM for simplicity. The system modeled in CONTAM is still referred to as a "VAV" system in the body of this text. Varying the supply flow rate can be implemented in CONTAM using controls and/or schedules by users who wish to do so. The supply air, return air, outside air, and exhaust flow rates modeled in CONTAM are listed in Table A32. In EnergyPlus, each HVAC system is assigned a minimum ventilation requirement and is modeled as such in CONTAM. For AHU 1, the minimum ventilation requirement is 1.82 m^3/s and 1.90 m^3/s for AHU 2 for all building vintages.

All of the exhaust flow rates (except for the "Toilet" zones) are modeled in EnergyPlus and CONTAM. The exhaust flow rates for the "Toilet" zones were modeled only in CONTAM, not in EnergyPlus. In CONTAM, the exhaust fans are modeled either as direct (on an exterior wall) exhaust fans or using ducts, depending on whether the room has an exterior wall.

The EnergyPlus model specifies a constant rate of air exchange between several zones that have no local exhaust. These zones also have no mechanical supply air. Thus, this is a method to drive air movement in the building. In CONTAM, this is done by pressurizing the building and allowing some transfer air to occur through natural driving forces and transfer grilles.

In EnergyPlus, neutral building pressurization is modeled in all zones. To pressurize the building in CONTAM, less air is returned than is supplied to each zone. For all building vintages, the return airflow rate is set to 90 % of the supply airflow rate. In CONTAM, only the MRI Room is neutrally pressurized. The return airflow rate is the supply airflow rate minus the exhaust rate for all building vintages.

Schedules:

The AHU 1 and first floor exhaust fans operate on the following schedule:
- Weekdays: 4:00 a.m. to 9:00 p.m., off otherwise
- Saturdays: 4:00 a.m. to 9:00 p.m., off otherwise
- Sundays and holidays: off all day

Outside air for AHU 1 is supplied according to this schedule as well.

The AHU 2 and second and third floor exhaust fans operate on the following schedule:
- Weekdays: 6:00 a.m. to 6:00 p.m., off otherwise
- Saturdays: 6:00 a.m. to 6:00 p.m., off otherwise
- Sundays and holidays: off all day

Outside air for AHU 2 is supplied according to this schedule as well.

Occupants:

The peak number of people for each zone is listed in Table A31. Occupants in all building zones are scheduled according to Figure A39. There is a different occupancy schedule for weekdays, Saturdays, and Sundays/ holidays.

Table A32 Summary of VAV system flow rates (m³/s) in Outpatient Healthcare

Zone	Floor	New		Post-1980		Pre-1980		Outside Air (m³/s)	Exhaust air (m³/s)
		Supply	Return	Supply	Return	Supply	Return		
Anesthesia	1	0.08	0.07	0.08	0.07	0.09	0.08	0.01	0.07
Bio Hazard	1	0.005	0.005	0.006	0.005	0.006	0.005	0.001	0
Café	1	0.43	0.39	0.46	0.41	0.49	0.44	0.07	0
Clean	1	0.06	0.05	0.06	0.06	0.06	0.06	0.01	0
Clean Work	1	0.08	0.07	0.08	0.07	0.08	0.07	0.01	0
Dictation	1	0.03	0.03	0.04	0.04	0.04	0.04	0.01	0
Dressing Room	1	0.04	0.04	0.07	0.06	0.08	0.07	0.01	0
Electrical Room	1	N/A	N/A	N/A	N/A	N/A	N/A	0	0
Elevator Pump Room	1	N/A	N/A	N/A	N/A	N/A	N/A	0	0
Humid	1	0.02	0.01	0.02	0.02	0.02	0.02	0.003	0
IT Hall	1	0.02	0.02	0.02	0.02	0.02	0.02	0.003	0
IT Room	1	0.03	0.03	0.04	0.03	0.04	0.03	0.01	0
Lobby	1	0.53	0.47	0.53	0.47	0.59	0.53	0.09	0
Lobby Hall	1	0.03	0.03	0.03	0.03	0.03	0.03	0.005	0
Lobby Toilet	1	N/A	N/A	N/A	N/A	N/A	N/A	0	0.04
Locker Room	1	0.35	0.32	0.38	0.34	0.39	0.35	0.06	0
Locker Room Hall	1	0.07	0.06	0.06	0.05	0.06	0.05	0.01	0
Lounge	1	0.18	0.16	0.18	0.17	0.18	0.17	0.03	0
Med Gas	1	0.05	0.05	0.09	0.08	0.11	0.10	0.01	0
MRI Control Room	1	0.08	0.07	0.10	0.09	0.10	0.09	0.01	0.08
MRI Hall	1	0.02	0.02	0.02	0.02	0.02	0.02	0.003	0
MRI Room	1	2.39	2.18	2.45	2.24	2.45	2.24	0.38	0.21
MRI Toilet	1	N/A	N/A	N/A	N/A	N/A	N/A	0	0.04
Nourishment	1	0.09	0.08	0.10	0.09	0.10	0.09	0.01	0
Nurse Hall	1	0.07	0.06	0.06	0.05	0.06	0.05	0.01	0
Nurse Janitor	1	N/A	N/A	N/A	N/A	N/A	N/A	0	0
Nurse Station	1	0.12	0.11	0.14	0.13	0.14	0.13	0.02	0
Nurse Toilet	1	N/A	N/A	N/A	N/A	N/A	N/A	0	0.04
Office	1	0.12	0.11	0.15	0.14	0.15	0.14	0.02	0
Operating Room 1	1	1.11	1.00	1.45	1.31	1.51	1.36	0.21	0
Operating Room 2	1	1.18	1.06	1.47	1.32	1.53	1.38	0.22	0
Operating Room 3	1	0.97	0.87	1.25	1.13	1.25	1.13	0.18	0
PACU	1	0.06	0.06	0.07	0.07	0.07	0.07	0.01	0
Pre-Op Hall	1	0.07	0.06	0.06	0.05	0.06	0.05	0.01	0
Pre-Op Room 1	1	0.09	0.08	0.09	0.08	0.10	0.09	0.01	0
Pre-Op Room 2	1	0.16	0.14	0.16	0.14	0.16	0.14	0.03	0
Pre-Op Toilet	1	N/A	N/A	N/A	N/A	N/A	N/A	0	0.04
Procedure Room	1	0.34	0.30	0.34	0.30	0.34	0.30	0.05	0
Reception	1	0.26	0.23	0.25	0.23	0.27	0.24	0.04	0
Reception Hall	1	0.02	0.02	0.02	0.01	0.02	0.01	0.002	0
Recovery Room	1	0.38	0.34	0.48	0.44	0.53	0.48	0.07	0
Scheduling	1	0.04	0.04	0.06	0.05	0.06	0.06	0.01	0
Scrub	1	0.012	0.011	0.010	0.009	0.010	0.009	0.002	0
Soil	1	0.10	0.09	0.10	0.09	0.10	0.09	0.02	0.10
Soil Hold	1	0.04	0.04	0.04	0.04	0.04	0.04	0.01	0.04

Zone	Floor	New		Post-1980		Pre-1980		Outside	Exhaust
Soil Work	1	0.14	0.13	0.14	0.13	0.14	0.13	0.02	0.14
Step Down	1	0.27	0.24	0.33	0.30	0.37	0.34	0.05	0
Sterile Hall	1	0.10	0.09	0.09	0.08	0.10	0.09	0.01	0
Sterile Storage	1	0.03	0.03	0.04	0.03	0.04	0.03	0.01	0
Storage	1	N/A	N/A	N/A	N/A	N/A	N/A	0	0
Sub-Sterile	1	0.09	0.08	0.12	0.10	0.13	0.12	0.02	0
Utility Hall	1	0.14	0.13	0.19	0.17	0.20	0.18	0.03	0
Utility Janitor	1	N/A	N/A	N/A	N/A	N/A	N/A	0	0
Utility Room	1	N/A	N/A	N/A	N/A	N/A	N/A	0	0
Vestibule	1	0.07	0.06	0.07	0.07	0.08	0.07	0.01	0
AHU 1 Total		**10.55**	**9.52**	**11.97**	**10.81**	**12.39**	**11.18**	**1.82**	
Conference	2	0.36	0.33	0.42	0.38	0.47	0.42	0.06	0
Conference Toilet	2	N/A	N/A	N/A	N/A	N/A	N/A	0	0.05
Dictation	2	0.029	0.026	0.034	0.031	0.034	0.031	0.005	0
Exam 1	2	0.44	0.40	0.48	0.44	0.56	0.50	0.07	0
Exam 2	2	0.33	0.30	0.35	0.32	0.38	0.34	0.05	0
Exam 3	2	0.44	0.40	0.48	0.43	0.51	0.46	0.07	0
Exam 4	2	0.07	0.06	0.08	0.07	0.09	0.08	0.01	0
Exam 5	2	0.33	0.30	0.34	0.31	0.39	0.35	0.05	0
Exam 6	2	0.16	0.14	0.18	0.16	0.19	0.17	0.03	0
Exam 7	2	0.45	0.41	0.49	0.44	0.50	0.45	0.07	0
Exam 8	2	0.18	0.16	0.20	0.18	0.21	0.19	0.03	0
Exam 9	2	0.23	0.21	0.25	0.22	0.25	0.23	0.04	0
Exam Hall 1	2	0.047	0.043	0.042	0.038	0.047	0.042	0.01	0
Exam Hall 2	2	0.047	0.043	0.042	0.038	0.047	0.042	0.01	0
Exam Hall 3	2	0.047	0.043	0.042	0.038	0.047	0.042	0.01	0
Exam Hall 4	2	0.051	0.046	0.045	0.040	0.044	0.040	0.01	0
Exam Hall 5	2	0.051	0.046	0.045	0.040	0.044	0.040	0.01	0
Exam Hall 6	2	0.051	0.046	0.044	0.040	0.044	0.040	0.01	0
Janitor	2	N/A	N/A	N/A	N/A	N/A	N/A	0	0
Lounge	2	0.06	0.05	0.06	0.05	0.06	0.05	0.01	0
Nurse Station 1	2	0.09	0.08	0.11	0.10	0.11	0.10	0.02	0
Nurse Station 2	2	0.11	0.10	0.13	0.12	0.13	0.12	0.02	0
Office	2	0.42	0.38	0.47	0.43	0.55	0.49	0.07	0
Office Hall	2	0.12	0.11	0.11	0.10	0.11	0.10	0.02	0
Reception	2	0.87	0.78	0.88	0.79	0.95	0.86	0.13	0
Reception Hall	2	0.24	0.22	0.22	0.20	0.35	0.31	0.04	0
Reception Toilet	2	N/A	N/A	N/A	N/A	N/A	N/A	0	0.10
Scheduling 1	2	0.12	0.11	0.15	0.13	0.15	0.13	0.02	0
Scheduling 2	2	0.13	0.12	0.15	0.14	0.15	0.14	0.02	0
Storage 1	2	N/A	N/A	N/A	N/A	N/A	N/A	0	0
Storage 2	2	N/A	N/A	N/A	N/A	N/A	N/A	0	0
Storage 3	2	N/A	N/A	N/A	N/A	N/A	N/A	0	0
Utility	2	N/A	N/A	N/A	N/A	N/A	N/A	0	0
Work	2	0.67	0.60	0.81	0.73	0.85	0.77	0.11	0
Work Hall	2	0.21	0.19	0.18	0.17	0.19	0.17	0.03	0
Work Toilet	2	N/A	N/A	N/A	N/A	N/A	N/A	0	0.04
X-Ray	2	0.43	0.38	0.54	0.48	0.54	0.48	0.07	0

Zone	Floor	New		Post-1980		Pre-1980		Outside	Exhaust
Dressing Room	3	0.022	0.020	0.025	0.022	0.024	0.022	0.003	0
Elevator Hall	3	0.15	0.14	0.16	0.14	0.18	0.17	0.02	0
Humid	3	0.050	0.045	0.057	0.051	0.062	0.056	0.01	0
Janitor	3	N/A	N/A	N/A	N/A	N/A	N/A	0	0
Locker	3	0.09	0.08	0.09	0.08	0.10	0.09	0.01	0
Lounge	3	0.62	0.56	0.62	0.56	0.68	0.62	0.09	0
Lounge Toilet	3	N/A	N/A	N/A	N/A	N/A	N/A	0	0.15
Mechanical	3	N/A	N/A	N/A	N/A	N/A	N/A	0	0
Mechanical Hall	3	0.11	0.10	0.11	0.10	0.11	0.10	0.02	0
Office	3	1.52	1.37	1.72	1.55	1.94	1.75	0.25	0
Office Hall	3	0.30	0.27	0.29	0.27	0.30	0.27	0.04	0
Office Toilet	3	N/A	N/A	N/A	N/A	N/A	N/A	0	0.04
Physical Therapy 1	3	1.09	0.98	1.22	1.10	1.40	1.26	0.18	0
Physical Therapy 2	3	0.36	0.33	0.39	0.35	0.42	0.38	0.06	0
Physical Therapy Toilet	3	N/A	N/A	N/A	N/A	N/A	N/A	0	0.07
Storage 1	3	N/A	N/A	N/A	N/A	N/A	N/A	0	0
Storage 2	3	N/A	N/A	N/A	N/A	N/A	N/A	0	0
Treatment	3	0.30	0.27	0.29	0.26	0.32	0.28	0.04	0
Undeveloped 1	3	N/A	N/A	N/A	N/A	N/A	N/A	0	0
Undeveloped 2	3	N/A	N/A	N/A	N/A	N/A	N/A	0	0
Utility	3	N/A	N/A	N/A	N/A	N/A	N/A	0	0
Work	3	0.44	0.39	0.49	0.44	0.58	0.52	0.07	0
AHU 2 Total		**11.86**	**10.67**	**12.84**	**11.56**	**14.08**	**12.67**	**1.90**	

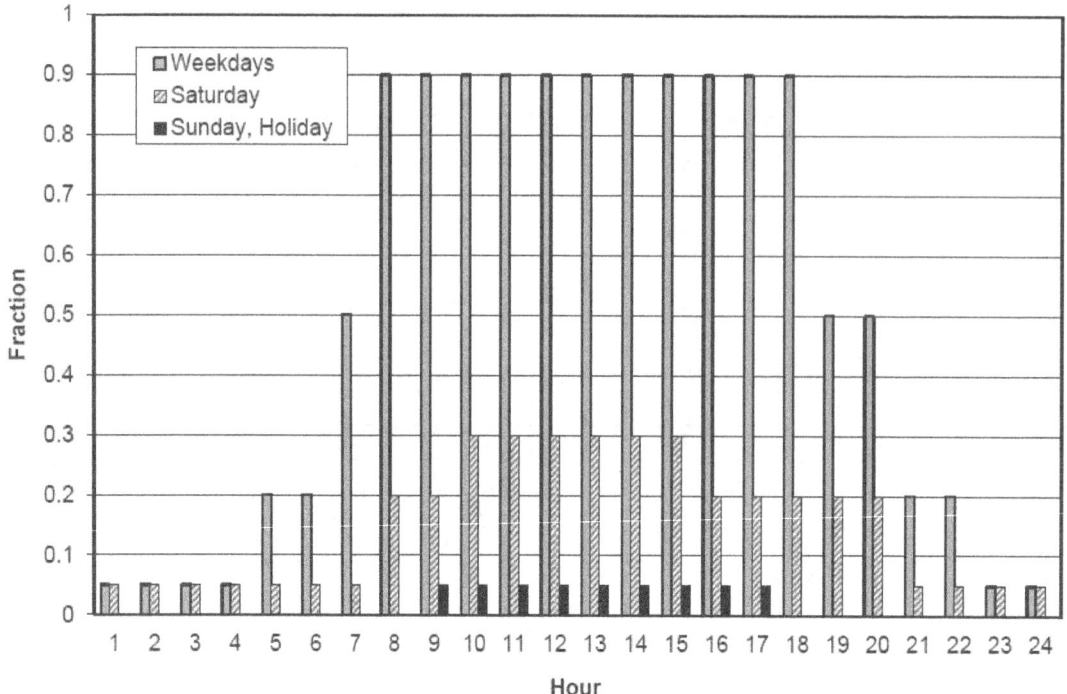

Figure A39 Occupancy schedule for Outpatient Health Care

A2.15. Warehouse

Table A33 summarizes the zones modeled in CONTAM for the Warehouse, their respective sizes, and maximum occupancy.

Table A33 Summary of zones in Warehouse

Zone	Area (m^2)	Height (m)	Maximum occupancy
Office	233	4.27	5
Fine Storage	1393	8.53	0
Bulk Storage	3205	8.53	0
Restroom	4	4.27	0

Geometry:

4598 m^2 footprint, one-story building with flat roof. The EnergyPlus model has three zones. Only the Office zone is 4.267 m high. The area above the Office is open to the Fine Storage, adding to the volume of that zone. The remaining zones are all 8.534 m high. In the CONTAM model, a Restroom (shaded in Figure A40) with a footprint of 2 m × 2 m was carved out of the Office.

134

Large interior leakage paths were defined as follows:

- Between Bulk and Fine Storage zones, a single large leakage path of 32 m^2 is modeled;
- Between Restroom and Office zones, a 0.025 m^2 door undercut is modeled.

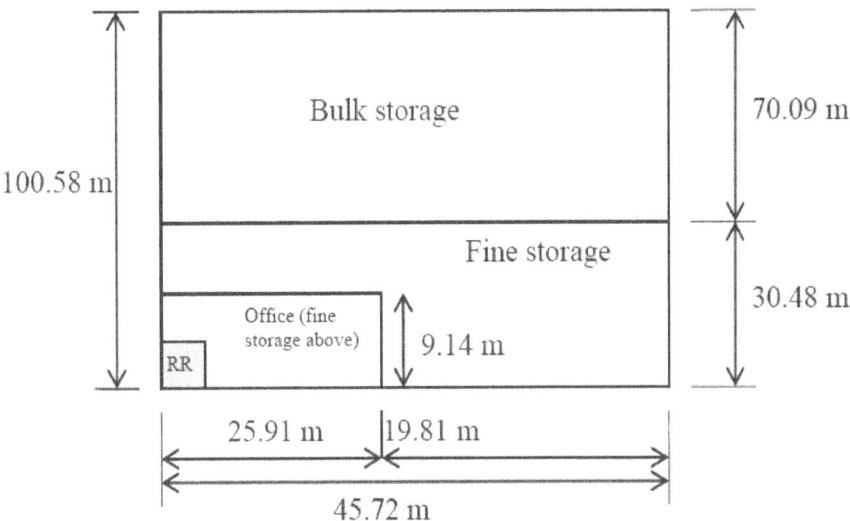

Figure A40 Floor plan of Warehouse (height 8.534 m, except for Office which is 4.267 m high)

HVAC systems:

For all building vintages, the EnergyPlus model has two CAV systems. One serves the Office and the other serves the Fine Storage zone. The supply air, return air, outside air, and exhaust flow rates modeled in CONTAM are listed in Table A34. The exhaust flow rate for the Restroom is modeled only in CONTAM, not in EnergyPlus.

The EnergyPlus model specifies a constant rate of air exchange in the Bulk Storage zone (0.00025 m^3/s/m^2 or 0.80 m^3/s). This zone also has no mechanical supply air. Thus, this is a method to drive air movement in the building. In CONTAM, a dedicated ventilation fan supplies 0.80 m^3/s of outdoor air. The Bulk Storage also has a unit heater in EnergyPlus that recirculates air locally within the zone and does not impact whole-building airflow or introduce outside air. Therefore, the unit heater is not modeled in CONTAM.

In EnergyPlus, neutral building pressurization is modeled in all zones. To pressurize the building in CONTAM, less air is returned than is supplied to each zone. For all building vintages, the return airflow rate is equal to the supply airflow rate minus the outside air requirement.

Table A34 Summary of CAV system flow rates (m³/s) in Warehouse

Zone	New		Post-1980		Pre-1980		Outside Air (m³/s)	Exhaust air (m³/s)
	Supply	Return	Supply	Return	Supply	Return		
Office	1.27	1.22	1.87	1.82	2.10	2.05	0.05	0
Fine Storage	5.03	4.68	8.68	8.33	9.29	8.94	0.35	0
Bulk Storage	N/A	N/A	N/A	N/A	N/A	N/A	0.80	0
Restroom	N/A	N/A	N/A	N/A	N/A	N/A	0	0.03

Schedules:

The HVAC and exhaust fans operate on the following schedule:
- Weekdays: 6:00 a.m. to 5:00 p.m., off otherwise
- Saturdays: 7:00 a.m. to 4:00 p.m., off otherwise
- Sundays and holidays: off all day

Outside air for AHU 1 is supplied according to this schedule as well.

Occupants:

The peak number of people for each zone is listed in Table A33. Occupants in all building zones are scheduled according to Figure A41. There is a different occupancy schedule for weekdays, and Saturdays. Sundays and holidays are unoccupied.

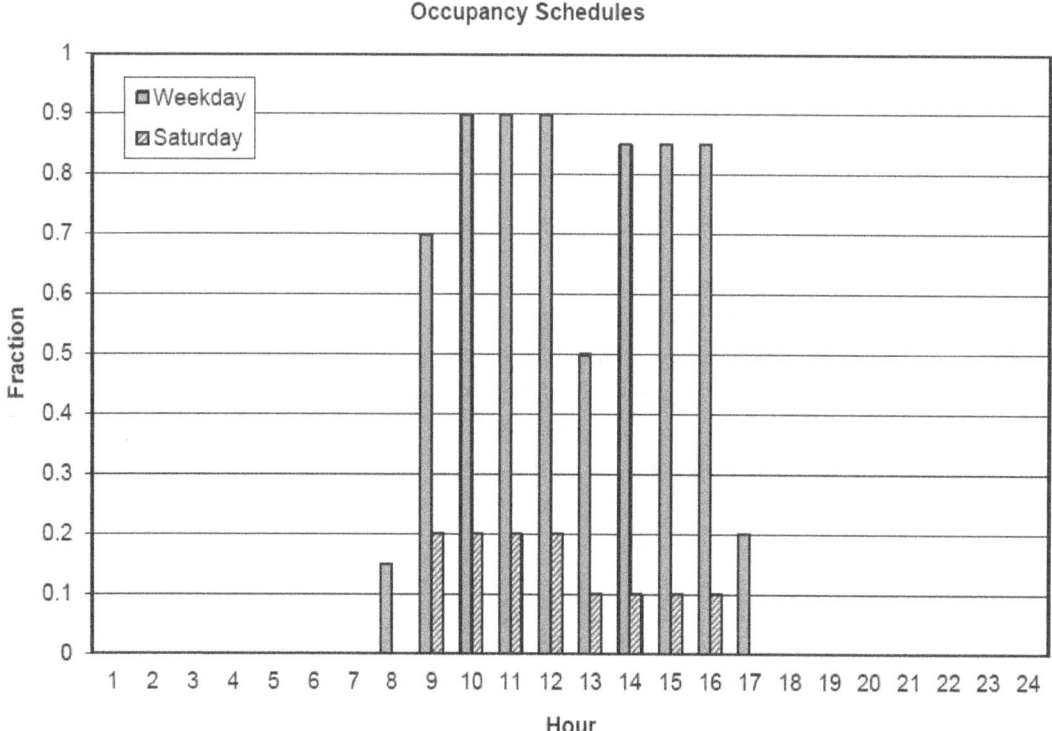

Figure A41 Occupancy schedule for Warehouse

136

A2.16. Midrise Apartment

Table A35 summarizes the zones modeled in CONTAM for the Warehouse, their respective sizes, maximum occupancy, and outside air rates.

Geometry:

784 m^2 footprint, four-story building with flat roof. The EnergyPlus model has 9 zones per floor. All floors have identical floor plans. In the CONTAM model, a Stairwell and Elevator Shaft (shaded in Figure A42) with footprints of 1.68 m × 4.0 m each were carved out of the Corridors.

Large interior leakage paths were defined as follows:
- A stairwell is defined using CONTAM's stair shaft model for closed treads and zero people;
- An elevator shaft is defined using CONTAM's elevator shaft model.

Table A35 Summary of zones and outside air rates in Midrise Apartment

Zone	Floor	Area (m^2)	Height (m)	Occupants	Outside Air (m^3/s)
G SW APT	1	88	3.05	2.50	0.042
G NW APT	1	88	3.05	2.50	0.042
OFFICE	1	88	3.05	2.00	0.020
G NE APT	1	88	3.05	2.50	0.042
G N1 APT	1	88	3.05	2.50	0.042
G N2 APT	1	88	3.05	2.50	0.042
G S1 APT	1	88	3.05	2.50	0.042
G S2 APT	1	88	3.05	2.50	0.042
M SW APT	2,3	88	3.05	2.50	0.042
M NW APT	2,3	88	3.05	2.50	0.042
M SE APT	2,3	88	3.05	2.50	0.042
M NE APT	2,3	88	3.05	2.50	0.042
M N1 APT	2,3	88	3.05	2.50	0.042
M N2 APT	2,3	88	3.05	2.50	0.042
M S1 APT	2,3	88	3.05	2.50	0.042
M S2 APT	2,3	88	3.05	2.50	0.042
T SW APT	4	88	3.05	2.50	0.042
T NW APT	4	88	3.05	2.50	0.042
T SE APT	4	88	3.05	2.50	0.042
T NE APT	4	88	3.05	2.50	0.042
T N1 APT	4	88	3.05	2.50	0.042
T N2 APT	4	88	3.05	2.50	0.042
T S1 APT	4	88	3.05	2.50	0.042
T S2 APT	4	88	3.05	2.50	0.042
T CORRIDOR	4	78	3.05	0	0
G CORRIDOR	1	78	3.05	0	0
M CORRIDOR	2,3	78	3.05	0	0

Note: The "G" means ground, "M" means middle, and "T" means top. "APT" is apartment.

HVAC systems:

For all building vintages, the EnergyPlus model has 24 packaged single-zone constant-volume systems, each serving a zone. In EnergyPlus, the systems do not supply outdoor air to the spaces, and there is no mixing of air between the zones. Thus, they are modeled as individual supply air fans in CONTAM for each zone. The rates are equal to the outside air flow rate listed in Table A35. There are no exhaust fans modeled in this building.

In EnergyPlus, the Corridors have a unit heater that recirculates air locally within the zone and does not impact whole-building airflow or introduce outside air. Therefore, they are not modeled in CONTAM.

Schedules:
All the HVAC system fans operate 24 hours per day every day of the year. Outside air is also supplied all of the time for the apartments. Outside air is supplied from 8:00 a.m. to 5:00 p.m. on weekdays for the Office. No outside air is supplied to the Office on weekends or holidays.

Occupants:
The peak number of people for each zone is listed in Table A35. Occupants in all building zones are scheduled according to Figure A43. There are two different occupancy schedules. One schedule for the apartments, and one for the Office. The Office is unoccupied on weekends and holidays.

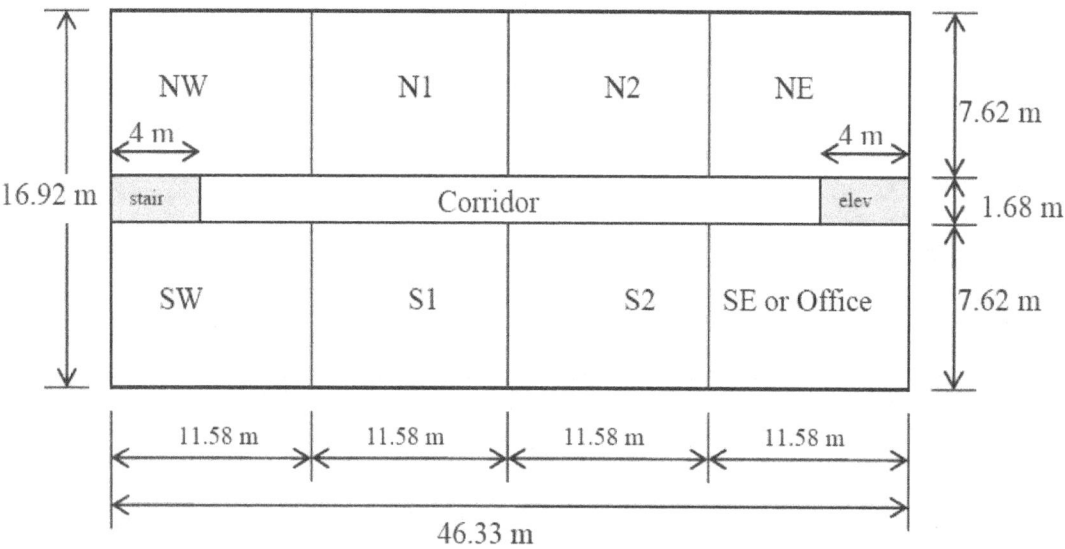

Figure A42 Floor plan of Midrise Apartment (height 3.05 m)

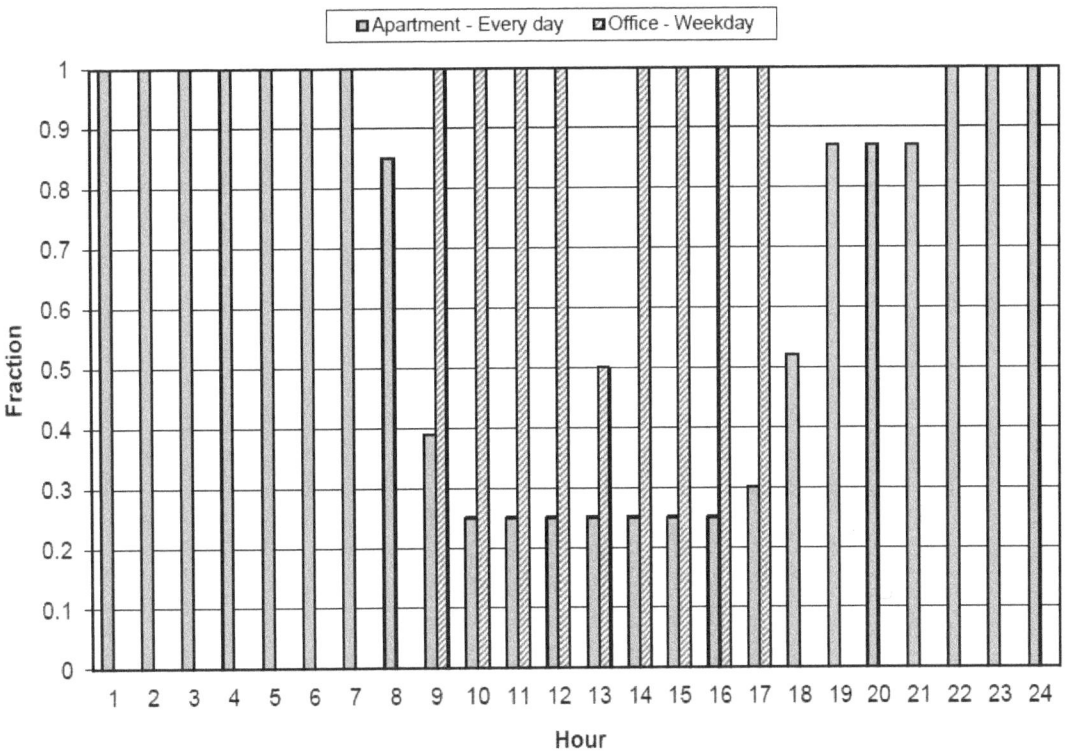

Figure A43 Occupancy schedules for Midrise Apartment

Appendix B Detailed calculated contaminant concentration predictions

B1. Full Service Restaurant

Dining	Daily average contaminant concentrations				Daily peak contaminant concentrations			
	Mean	Min.	Max.	StdDev	Mean	Min.	Max.	StdDev
CO_2, mg/m^3	1474	1235	1579	73	2210	1830	2433	114
Ozone, µg/m^3	23	3	53	10	37	5	75	13
PM 2.5, µg/m^3	10	1	32	5	15	2	44	7
VOC, µg/m^3	47	34	62	5	158	41	343	56

Kitchen	Daily average contaminant concentrations				Daily peak contaminant concentrations			
	Mean	Min.	Max.	StdDev	Mean	Min.	Max.	StdDev
CO_2, mg/m^3	1167	1020	1255	49	1606	1352	1762	82
Ozone, µg/m^3	28	4	62	12	45	7	86	16
PM 2.5, µg/m^3	13	1	42	7	21	2	61	10
VOC, µg/m^3	38	27	53	5	130	31	341	64

B2. Hospital

1F ER Exam 3	Daily average contaminant concentrations				Daily peak contaminant concentrations			
	Mean	Min.	Max.	StdDev	Mean	Min.	Max.	StdDev
CO_2, mg/m^3	975	919	991	13	1052	982	1070	16
Ozone, µg/m^3	17	2	37	7	28	4	54	10
PM 2.5, µg/m^3	7	1	24	4	12	1	33	6
VOC, µg/m^3	65	51	69	3	68	58	75	3

1F ER Nurse's Station	Daily average contaminant concentrations				Daily peak contaminant concentrations			
	Mean	Min.	Max.	StdDev	Mean	Min.	Max.	StdDev
CO_2, mg/m^3	862	764	908	52	946	781	1026	89
Ozone, µg/m^3	6	1	14	3	10	1	19	3
PM 2.5, µg/m^3	5	0	18	3	8	1	23	4
VOC, µg/m^3	198	123	224	24	214	169	226	11

1F Lobby	Daily average contaminant concentrations				Daily peak contaminant concentrations			
	Mean	Min.	Max.	StdDev	Mean	Min.	Max.	StdDev
CO_2, mg/m^3	860	751	914	55	947	765	1037	95
Ozone, µg/m^3	6	1	12	2	10	1	21	3
PM 2.5, µg/m^3	5	0	20	3	8	1	28	4
VOC, µg/m^3	206	108	242	30	225	148	243	19

2F ICU	Daily average contaminant concentrations				Daily peak contaminant concentrations			
	Mean	Min.	Max.	StdDev	Mean	Min.	Max.	StdDev
CO_2, mg/m^3	865	795	897	37	938	820	990	64
Ozone, µg/m^3	10	1	23	5	18	3	39	6
PM 2.5, µg/m^3	6	0	20	3	10	1	26	4
VOC, µg/m^3	119	84	132	11	127	106	134	5

2F ICU Patient Room 3	Daily average contaminant concentrations				Daily peak contaminant concentrations			
	Mean	Min.	Max.	StdDev	Mean	Min.	Max.	StdDev
CO_2, mg/m^3	877	780	922	46	960	805	1034	79
Ozone, µg/m^3	10	1	21	4	17	2	38	6
PM 2.5, µg/m^3	6	0	21	3	9	1	30	5
VOC, µg/m^3	175	88	202	24	190	118	203	14

2F Operating Room 2	Daily average contaminant concentrations				Daily peak contaminant concentrations			
	Mean	Min.	Max.	StdDev	Mean	Min.	Max.	StdDev
CO_2, mg/m^3	824	808	832	3	855	836	869	5
Ozone, µg/m^3	18	2	40	8	30	5	58	11
PM 2.5, µg/m^3	7	1	21	4	11	1	31	5
VOC, µg/m^3	47	38	55	3	49	42	60	4

(Hospital continued)

3F Lab	Daily average contaminant concentrations				Daily peak contaminant concentrations			
	Mean	Min.	Max.	StdDev	Mean	Min.	Max.	StdDev
CO_2, mg/m³	870	762	920	58	962	778	1046	101
Ozone, µg/m³	8	1	19	4	14	2	28	5
PM 2.5, µg/m³	6	0	19	3	9	1	24	4
VOC, µg/m³	178	105	205	23	192	141	206	14

3F Nurse's Station Lobby	Daily average contaminant concentrations				Daily peak contaminant concentrations			
	Mean	Min.	Max.	StdDev	Mean	Min.	Max.	StdDev
CO_2, mg/m³	842	749	881	48	918	759	982	82
Ozone, µg/m³	6	1	14	3	10	2	20	4
PM 2.5, µg/m³	5	0	17	3	8	1	22	4
VOC, µg/m³	154	110	171	17	166	146	173	5

3F Patient Room 3	Daily average contaminant concentrations				Daily peak contaminant concentrations			
	Mean	Min.	Max.	StdDev	Mean	Min.	Max.	StdDev
CO_2, mg/m³	884	799	926	42	966	827	1036	73
Ozone, µg/m³	9	1	24	4	15	2	36	6
PM 2.5, µg/m³	6	0	19	3	9	1	26	4
VOC, µg/m³	168	101	195	22	182	132	197	15

3F Patient Room 4	Daily average contaminant concentrations				Daily peak contaminant concentrations			
	Mean	Min.	Max.	StdDev	Mean	Min.	Max.	StdDev
CO_2, mg/m³	861	796	892	33	930	823	984	57
Ozone, µg/m³	11	1	26	5	19	3	40	7
PM 2.5, µg/m³	6	0	20	3	10	1	27	5
VOC, µg/m³	108	75	119	11	116	96	120	4

5F Dining	Daily average contaminant concentrations				Daily peak contaminant concentrations			
	Mean	Min.	Max.	StdDev	Mean	Min.	Max.	StdDev
CO_2, mg/m³	897	753	967	77	1016	768	1145	133
Ozone, µg/m³	8	1	20	4	14	2	33	5
PM 2.5, µg/m³	6	0	18	3	9	1	24	4
VOC, µg/m³	127	88	141	14	137	119	142	4

5F Office 2	Daily average contaminant concentrations				Daily peak contaminant concentrations			
	Mean	Min.	Max.	StdDev	Mean	Min.	Max.	StdDev
CO_2, mg/m³	860	751	904	54	948	763	1026	94
Ozone, µg/m³	10	1	24	4	18	3	37	7
PM 2.5, µg/m³	6	0	20	3	10	1	28	4
VOC, µg/m³	116	73	129	13	126	101	130	4

B3. Medium Office

1F Core Zone	Daily average contaminant concentrations				Daily peak contaminant concentrations			
	Mean	Min.	Max.	StdDev	Mean	Min.	Max.	StdDev
CO_2, mg/m^3	1111	876	1219	95	1289	966	1409	126
Ozone, µg/m^3	8	2	18	3	12	2	26	4
PM 2.5, µg/m^3	5	0	18	3	7	1	24	4
VOC, µg/m^3	192	120	287	36	331	131	812	152

2F Core Zone	Daily average contaminant concentrations				Daily peak contaminant concentrations			
	Mean	Min.	Max.	StdDev	Mean	Min.	Max.	StdDev
CO_2, mg/m^3	1110	883	1207	92	1285	977	1416	121
Ozone, µg/m^3	8	1	18	3	12	2	26	4
PM 2.5, µg/m^3	5	0	16	3	7	1	23	4
VOC, µg/m^3	190	124	259	28	319	149	685	117

3F Core Zone	Daily average contaminant concentrations				Daily peak contaminant concentrations			
	Mean	Min.	Max.	StdDev	Mean	Min.	Max.	StdDev
CO_2, mg/m^3	1093	882	1201	87	1261	975	1414	117
Ozone, µg/m^3	8	1	19	4	12	2	27	5
PM 2.5, µg/m^3	5	0	16	3	7	1	22	4
VOC, µg/m^3	181	122	225	20	294	149	505	69

1F South Zone	Daily average contaminant concentrations				Daily peak contaminant concentrations			
	Mean	Min.	Max.	StdDev	Mean	Min.	Max.	StdDev
CO_2, mg/m^3	1100	832	1216	100	1278	896	1405	132
Ozone, µg/m^3	8	1	28	4	13	2	52	6
PM 2.5, µg/m^3	5	0	19	3	8	1	29	5
VOC, µg/m^3	190	75	291	41	297	106	812	143

2F South Zone	Daily average contaminant concentrations				Daily peak contaminant concentrations			
	Mean	Min.	Max.	StdDev	Mean	Min.	Max.	StdDev
CO_2, mg/m^3	1081	841	1183	92	1248	909	1383	121
Ozone, µg/m^3	9	2	30	4	14	2	54	6
PM 2.5, µg/m^3	5	0	16	3	7	1	24	4
VOC, µg/m^3	181	77	254	31	291	110	659	111

3F South Zone	Daily average contaminant concentrations				Daily peak contaminant concentrations			
	Mean	Min.	Max.	StdDev	Mean	Min.	Max.	StdDev
CO_2, mg/m^3	1058	842	1172	86	1215	909	1374	116
Ozone, µg/m^3	10	2	34	5	15	2	57	7
PM 2.5, µg/m^3	5	0	16	3	8	1	24	4
VOC, µg/m^3	169	77	217	24	270	109	486	68

1F West Zone	Daily average contaminant concentrations				Daily peak contaminant concentrations			
	Mean	Min.	Max.	StdDev	Mean	Min.	Max.	StdDev
CO_2, mg/m^3	1055	826	1161	88	1209	887	1328	117
Ozone, µg/m^3	11	2	25	5	17	2	42	6
PM 2.5, µg/m^3	5	0	23	3	8	1	32	5
VOC, µg/m^3	170	83	263	38	294	104	720	143

2F West Zone	Daily average contaminant concentrations				Daily peak contaminant concentrations			
	Mean	Min.	Max.	StdDev	Mean	Min.	Max.	StdDev
CO_2, mg/m^3	1049	833	1142	82	1200	898	1328	110
Ozone, µg/m^3	11	2	26	5	17	2	42	7
PM 2.5, µg/m^3	5	0	21	3	8	1	29	4
VOC, µg/m^3	166	88	233	29	288	115	594	109

3F West Zone	Daily average contaminant concentrations				Daily peak contaminant concentrations			
	Mean	Min.	Max.	StdDev	Mean	Min.	Max.	StdDev
CO_2, mg/m^3	1030	834	1136	76	1173	900	1321	105
Ozone, µg/m^3	12	2	28	5	18	3	42	8
PM 2.5, µg/m^3	5	0	20	3	8	1	29	4
VOC, µg/m^3	156	89	195	21	263	116	472	63

B4. Primary School

Cafeteria	Daily average contaminant concentrations				Daily peak contaminant concentrations			
	Mean	Min.	Max.	StdDev	Mean	Min.	Max.	StdDev
CO_2, mg/m^3	859	801	878	24	1429	926	1572	226
Ozone, $\mu g/m^3$	34	6	74	15	50	14	98	19
PM 2.5, $\mu g/m^3$	16	1	52	9	24	2	68	12
VOC, $\mu g/m^3$	30	15	83	9	215	38	913	112

Gym	Daily average contaminant concentrations				Daily peak contaminant concentrations			
	Mean	Min.	Max.	StdDev	Mean	Min.	Max.	StdDev
CO_2, mg/m^3	1072	940	1143	50	1444	1008	1579	191
Ozone, $\mu g/m^3$	23	4	48	10	33	9	65	12
PM 2.5, $\mu g/m^3$	14	1	46	8	21	2	61	11
VOC, $\mu g/m^3$	70	42	146	14	364	80	1186	147

Library/Media Classroom	Daily average contaminant concentrations				Daily peak contaminant concentrations			
	Mean	Min.	Max.	StdDev	Mean	Min.	Max.	StdDev
CO_2, mg/m^3	1137	834	1279	131	1440	873	1683	251
Ozone, $\mu g/m^3$	14	3	29	6	20	5	41	7
PM 2.5, $\mu g/m^3$	8	0	30	5	13	1	39	7
VOC, $\mu g/m^3$	99	60	158	19	235	65	773	117

Offices	Daily average contaminant concentrations				Daily peak contaminant concentrations			
	Mean	Min.	Max.	StdDev	Mean	Min.	Max.	StdDev
CO_2, mg/m^3	961	816	1056	65	1130	869	1284	117
Ozone, $\mu g/m^3$	12	2	31	5	18	5	44	7
PM 2.5, $\mu g/m^3$	8	0	28	5	12	1	35	6
VOC, $\mu g/m^3$	105	64	172	17	235	76	780	102

Pod 1 Corner Classroom 1	Daily average contaminant concentrations				Daily peak contaminant concentrations			
	Mean	Min.	Max.	StdDev	Mean	Min.	Max.	StdDev
CO_2, mg/m^3	977	793	1068	83	1168	820	1325	159
Ozone, $\mu g/m^3$	24	4	55	10	35	9	73	13
PM 2.5, $\mu g/m^3$	10	1	36	6	16	1	47	8
VOC, $\mu g/m^3$	49	29	82	10	127	34	425	68

Pod 1 Multiple Classroom 1	Daily average contaminant concentrations				Daily peak contaminant concentrations			
	Mean	Min.	Max.	StdDev	Mean	Min.	Max.	StdDev
CO_2, mg/m^3	1097	821	1226	122	1381	856	1597	236
Ozone, $\mu g/m^3$	17	3	44	7	25	7	63	10
PM 2.5, $\mu g/m^3$	10	0	31	6	14	1	41	7
VOC, $\mu g/m^3$	73	35	125	15	208	42	715	105

B5. Small Hotel

Front Lounge	Daily average contaminant concentrations				Daily peak contaminant concentrations			
	Mean	Min.	Max.	StdDev	Mean	Min.	Max.	StdDev
CO_2, mg/m^3	824	759	870	35	1156	847	1363	185
Ozone, μg/m^3	20	3	43	8	33	5	70	11
PM 2.5, μg/m^3	11	1	36	6	18	2	47	8
VOC, μg/m^3	40	28	60	6	53	35	77	9

Meeting Room	Daily average contaminant concentrations				Daily peak contaminant concentrations			
	Mean	Min.	Max.	StdDev	Mean	Min.	Max.	StdDev
CO_2, mg/m^3	839	819	847	6	1151	1072	1169	17
Ozone, μg/m^3	27	3	61	12	45	7	89	16
PM 2.5, μg/m^3	13	1	41	7	21	2	60	9
VOC, μg/m^3	19	16	22	1	26	22	32	1

Guest 209-212	Daily average contaminant concentrations				Daily peak contaminant concentrations			
	Mean	Min.	Max.	StdDev	Mean	Min.	Max.	StdDev
CO_2, mg/m^3	941	819	1044	45	1050	900	1242	73
Ozone, μg/m^3	8	1	25	3	14	2	47	6
PM 2.5, μg/m^3	9	1	32	5	15	1	40	7
VOC, μg/m^3	181	79	278	43	207	103	297	50

Guest 309-312	Daily average contaminant concentrations				Daily peak contaminant concentrations			
	Mean	Min.	Max.	StdDev	Mean	Min.	Max.	StdDev
CO_2, mg/m^3	966	824	1054	43	1110	920	1257	68
Ozone, μg/m^3	7	1	25	3	13	1	48	6
PM 2.5, μg/m^3	9	1	29	5	14	1	39	7
VOC, μg/m^3	206	86	282	39	232	115	300	39

Guest 409-412	Daily average contaminant concentrations				Daily peak contaminant concentrations			
	Mean	Min.	Max.	StdDev	Mean	Min.	Max.	StdDev
CO_2, mg/m^3	973	821	1049	40	1134	918	1261	61
Ozone, μg/m^3	7	1	26	4	14	2	51	7
PM 2.5, μg/m^3	9	1	27	5	14	1	39	7
VOC, μg/m^3	212	85	281	35	242	115	297	32

Guest 215-218	Daily average contaminant concentrations				Daily peak contaminant concentrations			
	Mean	Min.	Max.	StdDev	Mean	Min.	Max.	StdDev
CO_2, mg/m^3	895	822	947	29	965	864	1043	41
Ozone, μg/m^3	8	1	20	3	14	2	34	5
PM 2.5, μg/m^3	8	1	27	4	12	1	38	6
VOC, μg/m^3	155	89	209	28	179	104	238	30

(Small Hotel continued)

Guest 315-318	Daily average contaminant concentrations				Daily peak contaminant concentrations			
	Mean	Min.	Max.	StdDev	Mean	Min.	Max.	StdDev
CO_2, mg/m^3	955	840	1041	42	1085	917	1239	64
Ozone, µg/m^3	7	1	19	3	13	2	33	6
PM 2.5, µg/m^3	8	1	28	5	13	1	41	6
VOC, µg/m^3	204	106	279	38	233	123	297	39

Guest 415-418	Daily average contaminant concentrations				Daily peak contaminant concentrations			
	Mean	Min.	Max.	StdDev	Mean	Min.	Max.	StdDev
CO_2, mg/m^3	975	842	1048	40	1134	933	1268	61
Ozone, µg/m^3	7	1	20	4	14	2	35	7
PM 2.5, µg/m^3	9	1	28	5	13	1	42	7
VOC, µg/m^3	217	109	280	35	249	130	306	32

B6. Stand-Alone Retail

Back Space	Daily average contaminant concentrations				Daily peak contaminant concentrations			
	Mean	Min.	Max.	StdDev	Mean	Min.	Max.	StdDev
CO_2, mg/m^3	846	741	1029	70	949	757	1217	121
Ozone, μg/m^3	10	1	29	5	15	2	43	7
PM 2.5, μg/m^3	7	0	32	5	11	1	45	7
VOC, μg/m^3	96	30	180	34	135	38	436	73

Core Retail	Daily average contaminant concentrations				Daily peak contaminant concentrations			
	Mean	Min.	Max.	StdDev	Mean	Min.	Max.	StdDev
CO_2, mg/m^3	1052	865	1173	76	1283	975	1476	123
Ozone, μg/m^3	10	2	22	4	15	2	31	5
PM 2.5, μg/m^3	8	0	28	5	11	1	36	6
VOC, μg/m^3	92	59	144	16	197	69	427	79

Front Retail	Daily average contaminant concentrations				Daily peak contaminant concentrations			
	Mean	Min.	Max.	StdDev	Mean	Min.	Max.	StdDev
CO_2, mg/m^3	988	794	1158	88	1195	845	1467	152
Ozone, μg/m^3	13	2	43	6	20	2	67	9
PM 2.5, μg/m^3	9	0	31	5	13	1	45	7
VOC, μg/m^3	64	20	128	21	97	25	421	63

Point of Sale	Daily average contaminant concentrations				Daily peak contaminant concentrations			
	Mean	Min.	Max.	StdDev	Mean	Min.	Max.	StdDev
CO_2, mg/m^3	984	794	1163	92	1190	843	1486	159
Ozone, μg/m^3	13	2	43	6	20	2	68	9
PM 2.5, μg/m^3	8	0	29	5	13	1	44	7
VOC, μg/m^3	62	20	126	22	93	24	358	59